NORTHERN BEE BOOKS
Scout Bottom Farm, Mytholmroyd, West Yorkshire
www.northernbeebooks.co.uk

BEESWAX by Ron Brown

ISBN: 978-1-908904-83-6

Published by Northern Bee Books, 2015
Scout Bottom Farm
Mytholmroyd
Hebden Bridge HX7 5JS (UK)

www.northernbeebooks.co.uk
Tel: 01422 882751

Printed by Lightning Source, UK

BEESWAX

'The honeybags steal from the humble-bees
and for wax tapers crop their waxen thighs,
and light them at the fiery glow-worms' eyes,
to have my love to bed, and to arise.'

Titania, in *A Midsummer Night's Dream*, Act
III, Scene i.
William Shakespeare.

PL. 1   The Normansell Cup, 1683

BEESWAX

Ron Brown O.B.E., B.Sc.

Former Editor of *Beekeeping* and author
of *A Simple Two-Queen System*
One Thousand Years of Devon Beekeeping
Beekeeping—A Seasonal Guide
Honey Bees—A Guide to Management
All Round the Compass
Great Masters of Beekeeping

NORTHERN BEE BOOKS
Scout Bottom Farm, Mytholmroyd, West Yorkshire
www.northernbeebooks.co.uk

2nd edition, limp,
revised 1989
ISBN 0 905652 150

3rd edition, 1995
ISBN 0 905652 363

4th edition, 2009
ISBN 978-0-0905652-36-8

4th edition, reprint, 2015
ISBN: 978-1-908904-83-6

The First Edition of this book was awarded the Silver Medal of Apimondia at the XXVIII International Congress held in Acapulco, Mexico, from 23rd to 29th October 1981.

Apimondia is, of course, the World Bee Organisation which holds an International Congress once every two years. These are attended by upwards of 2,000 delegates from almost every country in the world.

CONTENTS

FIG. 1. Chinoiserie engraving on the Normansell Cup, 1683

LIST OF ILLUSTRATIONS

ACKNOWLEDGEMENTS

I wish to express my gratitude to all those who have helped in the production of this book, and my apologies to any who may feel that a more specific acknowledgement would have been appropriate. When one has been reading books and journals, attending lectures and conferences as well as learning from the bees themselves for half a life-time, it is hard to recall the original source of much of the information which has become part of oneself over the years. Much of the historical material has come from the Worshipful Company of Wax Chandlers via the archives of the Guildhall Library, London, and especial tribute is paid to John Dummelow's history *The Wax Chandlers of London*, and to Mr A. G. Horton, Hon. Publications Secretary of the Central Association of Beekeepers.

Days spent with the Case-Green family at Redhill (The British Wax Refining Co.), Messrs Robert Lee of Uxbridge, at the Guildhall and other libraries, are remembered with pleasure and acknowledged with sincere appreciation, as are personal communications from Mrs Clara Furness on candle-making, Mr C. Brian Dennis on wax-handling, Mr Basil Salter on preparing wax for show, and others. Acknowledgement of happy weeks spent with skep beekeeper Mr John Starling of Burrough Green in 1934, when some practical uses of beeswax were first learned, and to Mr H. Howard (Uncle Bert) of Brinkley who first showed me wild combs from a hollow tree in 1919, must serve to cover the unrecorded scores of bee men to whom I remain indebted. My thanks to John Kinross, of Bee Books New and Old, who commissioned this work, to Karl Showler of I.B.R.A. who read the proofs, to the Director and staff of I.B.R.A. who went to great trouble to let me have a sight of original theses and papers of recent work.

R.H.B.

20 Parkhurst Road,
Torquay, Devon.
19 May 81

FEEDBACK FROM READERS

Over the years since 'Beeswax' was first published in 1981, I have received hundreds of queries and comments, by phone, letter, or at meetings. For the most part, they have dealt with recipes of one kind or another, including several expressing doubt about the wisdom of using liquid paraffin in cosmetics, alleging that such a 'mineral oil' could be carcinogenic. On this point I have tried to explain that 'paraffin' was a technical term for a class of hydrocarbons, from methane (a gas) to vaseline and some waxes. In this case 'liquid paraffin' was another name for 'medicinal paraffin', definitely not carcinogenic. Several correspondents suggested that I should have warned against melting wax in iron or copper vessels. They were right; I should have explained the discolouration which would arise. Stainless steel is best, but in practice, enamel, aluminium or even bright tin-plate have been used with no problems.

An entirely different query came from a keen bowls player in Tasmania, asking for a polish recipe which would give his woods a smooth surface but just a hint of tackiness. I suggested the incorporation of a very small amount of powdered resin into the normal recipe.

As one might expect, some brickbats were received, but also the occasional bouquet. I still cherish the memory of a beginner in Sussex, who came up after a lecture on queen-rearing, to thank me for the detailed instruction on how to produce a cake of wax. He said that he had followed every detail exactly and his very first cake of wax won First Prize in a show.

Torquay   R.H.B.
1995

CHAPTER ONE

BEESWAX AND HISTORY

By 3000 B.C. there was already an established technology involving the use of large quantities of wax. The ancient Egyptians used wax for embalming and modelling; exquisite examples of cats, eagles and other waxen figures have been recovered from graves and from the pyramids. Among the better known are a model head of the god Osiris and replicas of the scarabeus, or sacred beetle.

CLASSICAL TIMES

Catullus spoke of the ancient Roman love letters system by exchange of wax tablets, when the recipient rubbed out the message and wrote the reply on the smoothed sheet of wax. How sensible, with no evidence left behind for blackmail or breach of promise actions! A normal Roman writing set was the diptycha and stylus,

FIG. 2.   Wax writing tablet, stylus and casting

the former being a pair of facing light wooden pages thinly coated with beeswax and hinged together. The stylus was often made of carved ivory, with a sharp point for writing at one end and a broad chisel shape at the other for rubbing out.

Mural decorations worked in coloured beeswax adorned the

11

homes of wealthy Romans, and some fine specimens may be seen at Herculaneum, uncovered after having been buried in volcanic mud for 1800 years.

The pagan gods of ancient Rome, as well as Buddha in Ceylon, had candles lit in their honour, and the Christian use of candles was and is but part of a much older tradition. Herodotus wrote of the ancient Persians coating the bodies of important people with beeswax before burial, and in more recent times in England the dead were buried in cerecloths,* or material coated with wax, rather than in wooden coffins.

As a durable commodity of great practical use, never available in such quantities as might cheapen its value, beeswax was treated in the past almost as a currency like gold or silver. When the Romans conquered Corsica in 181 B.C. they imposed an annual tribute payable in beeswax. In Britain in 883 A.D. Aethelred Alderman was recorded as having to pay to the Church 'One sestar of honey and one sestar of wax' (Brit. Museum 2081a). A sestar was one-sixth of a gallon, so a sestar of wax would have weighed just over $1\frac{1}{2}$ lb.

Even though abbeys and monasteries usually kept bees themselves, their tithes and rents were commonly assessed at fixed quantities of wax per annum. For hundreds of years the price of wax remained almost constant at around 12 old pence (5p) per lb, about 8 times the price of honey. This despite inflation at the end of the 16th century and again during the Civil War. It was the introduction of paraffin wax candles and then oil lamps which reduced the demand for beeswax candles and brought down the price of wax.

The use of sealing wax also goes back at least to Roman times, and in those times a mixture of beeswax with a little resin was found to take a sharp impression of a signet ring, and to be more durable than pure wax. The clear pattern imposed by a unique

---

* 'Sincere', a word which in one sense can be applied to most beekeepers, comes from the Latin 'sine cere' meaning 'without wax'. The reference was to the habit of unscrupulous sculptors in ancient Roman times of using bleached beeswax to repair marble statues. Perhaps only after the statue had been paid for and taken away (and stood in the hot sun!), was the repair work noticed. So a statue guaranteed 'sine cere' came to be an example of honest good faith, and in time was applied to all dealings, and to people who could be trusted.

mould used for centuries as a guarantee of authenticity and proof that the document expressed the will of the owner of the seal. The signet or seal was jealously guarded, or worn on the person, to obviate fraud and unauthorised use. In times when most people could not write, the impression of their seal was the equivalent of a signature. Legal documents were thus sealed in Roman times, and the Great Seal of England is a fine example of its use to give the Royal Assent.

BEESWAX CANDLES AND THE CHURCH

Of all the many uses of wax over the centuries, perhaps the greatest has been in candles. From the 4th century the Christian Church has attached a special significance to candles made of beeswax. To this day in the Catholic Church the Pascal candle is blessed by the deacon before Mass on Holy Saturday, and represents the risen Christ. Hence it is lighted at liturgical functions during the 40 days between Easter and the Feast of the Ascension, and finally extinguished after the Gospel of this Feast. The altar candles must still have a content of beeswax, no longer 100% but varying from 51% to 5% by local episcopal instruction.

THE WAX CHANDLERS

Though the early wax chandlers left few records, various legal documents and city archives dating from the thirteenth century make mention of the craft of candle-making. Among the earliest names were Robert le Cierger in 1277 and William le Sirger in 1311. By 1330 a recognised 'mistery' or trade craft must have been in existence as the chandlers were 'invited' to contribute a 'present' to King Edward III to help pay for the French wars. They gave 40 shillings, which would equate to £980 today. Similar contribution were made in 1355 and again in 1363. The first evidence on record of any self-governing control was in 1343 when four 'cirgiarii' were sworn before the maire and authorised to investigate the adulteration of beeswax with fat. About 15 years later a series of ordinances were issued under the authority of Maire John de Stodeye (a wellknown vintner) and his aldermen. Amongst other things these forbad 'the inclusion in cierges (candles) of any fat, rosin or other manner of refuse; nor should old wax and worse be put within and new waxe without. Nor should wykes be of excessive weight, so as to be selling wyke for waxe, in deceit of the common people.' Today, over 600 years later, such an ordinance from Brussels would not surprise us. Human nature does not change!

In 1371 a further ordinance set out in greater detail several regulations to control the making and sale of candles, and quoted a firm price of 9d. per lb for beeswax. This would be equivalent to about £17 today.* The most important landmark was the Royal Charter of 1484 whereby King Richard III gave official recognition and powers to the 'mistery or crafte of Wexe Chaundelers of our Citie of London'.

Of the 92 Livery Companies of the City of London, with crafts and trading interests as widely differing as those of the Goldsmiths' and the Drapers', the Wax Chandlers ranks twentieth in order of precedence. Twice in their long history, the Wax Chandlers have contributed a Lord Mayor of London; both were knighted by the Sovereign.

In mediaeval times artificial light, apart from that given by log fires, was by candles of one sort or another. The very poor had none at all, except for cheap tallow candles made from mutton fat, burning with a smoky flame and an unpleasant smell. Those less poor had rushlights, made by dipping thin hollow reeds in molten beeswax, when the central pith of the reed stem absorbed the wax and also acted as a wick. The well-to-do had beeswax candles and the nobility and very wealthy had branched candelabra burning three, five or even more candles at a time, giving a good light and a pleasant aroma.

Thus the Wax Chandlers had a virtual monopoly of artificial light, one might say like the Central Electricity Generating Board

Inflation and the Value of Money*

1330	£1	1609	£4.70	1793	£ 8.75	1958	£35.00
1484	£1.07	1642	£7.00	1854	£10.00	1972	£70.00
1500	£0.75	1663	£7.00	1914	£ 9.00	1982	£272.00
1540	£1.40	1740	£7.60	1938	£15.00	1994	£490.00

*Figures from 1330–1972 taken from 'Wax Chandlers of London' by John Dummelow. Figures from 1972–1994 taken from Income Tax Indexation Tables.

does today. Their business was to process and sell beeswax, to make and sell candles and other wax products. This business went on virtually unchanged for centuries, until the introduction of paraffin wax and oil lamps in the 19th century. I have a vivid memory of childhood life in a Cambridgeshire village in the early 1920s, when oil lamps and candles gave the only light, when candlesticks were on the hall table to be lit at bedtime and we children went up to bed in cold bedrooms by flickering candle light (and read surreptitiously). These were paraffin wax candles, of course, but much cheaper tallow candles were still on sale at the same village shop, which had stocked them for a century or more. In our family they were only used for rubbing chests when we had stubborn coughs or colds, but I remember that they had wicks in them, and gave a smelly, smoky light. No gas, but electricity came to the village in the 1940's.

In past centuries the production of wax candles was big business, not only for everyday use but on state occasions, and for churches.

However the Reformation was getting under way, and what would be termed today the 'Church candle business' suffered accordingly, as Rood screens were destroyed, altars simplified, and monasteries taken over in the 1530s and 1540s.

At one stage in this period wax candles in church were forbidden, except for necessary provision of light, i.e. not allowed on altars. A few years later the last measure must have lapsed, for in 1588 a gigantic paschal candle weighing 300 lb was set up at the newly restored shrine of Edward the Confessor in Westminster Abbey, by the Wax Chandlers' Company. To quote verbatim from their records 'The xxi day of Marche ... Master and Wardens of the Wax Chandlers mad ther iii C of wax with xx more at the making, after a grett dener'.

The records show that in 1563, skilled tradesmen employed by Wax Chandlers were paid £2 13s 4d a year (with meat and drink provided), or 4s 6d a week without meat and drink.

A few years later the Company had been able to get a Regulating Act through Parliament, 23 Eliz. c. 8 of 1581 'An Act for the true melting, making and working in Waxe', with the interesting preamble: 'Whereas by the Goodness of God this land doth yield great plenty of honey and waxe not only for home consumption byt also for export ...'

Later on, Cromwell was also bad for business, and his edicts forbade the use of altar candles, wax effigies of persons and angels, etc., so that a lucrative church trade was once again curtailed. Just before this a number of accounts quoted prices for such items as 'iv effigies, iv angels, pins and sticks', with labour charged at 1d per lb of wax used (if old wax was supplied by the customer), otherwise 8d a lb for new wax.

On 4 September 1666 came the Great Fire of London, destroying 13,000 houses and 87 churches, St Paul's Cathedral among them; 44 Company Halls, including that of the Wax Chandlers, were also destroyed by the fire. The Hall was rebuilt within a few years.

This same Hall was burnt out in 1940 by enemy bombing, but many treasures were saved by the brave and devoted action of the Clerk, Albert Wood, who lived with his family in a flat at the top of the building. Many of the pictures and the best of the silver (including the Normansell Cup, hallmarked in 1683) were saved, but table silver, silk banners and a great deal more were destroyed, as were the lead cistern and piping, which were melted by the heat.

The high altar in St Paul's Cathedral was also destroyed in 1940. A new altar was consecrated in 1958 and the Company of Wax Chandlers undertook to maintain a supply of candles in perpetuity.

The Hall was rebuilt in 1958, still in Gresham Street, and is in use today. By long custom, Lord Mayors still lunch at the Wax Chandlers' Hall in January after their election, and by an equally pleasing though newer custom, the Central Association of the British Beekeepers' Association holds one lecture each year in the Hall. Among their many good works (supporting two pupils at the Bluecoat School, grants and donations to hospitals, the new City University and selected charities) is the annual award of the Wax Chandlers' Prize (£100) to the best candidate passing the B.B.K.A. Senior Examination. The link between beekeeping and the Wax Chandlers remains strong, with the Master ex officio patron of the B.B.K.A. Central Association.*

* The current Master, Dr Harry Riches is a prominent beekeeper and past President of the British Beekeepers' Association.

CHAPTER TWO

THE ORIGIN OF BEESWAX

Although beeswax has been a familiar article of commerce for three or four thousand years, it is only in the last two hundred years or so that anyone has had a clear idea of its origin, apart from the fact that bees were responsible for it.

WAX SCALES

The first serious study was made by John Hunter and recorded by him in a paper read to the Royal Society on 23 February 1792. Francis Huber carried out some very thorough research work in 1793, which he recorded in detail in chapter VI of his *New Observations*. He described exactly the four pairs of wax membranes 'of irregular pentagonal shape', and realised that a liquid diffused through the membranes on which the wax was moulded.

With the covering segments removed, and seen under a strong magnifying glass, each of the eight wax 'mirrors' appears as a shallow, tray-like mould, with a base of transparent membrane, porous to liquid secretions from wax glands beneath and framed by narrow edging of hard chitin. These are arranged in four pairs, each looking rather like exotic spectacles frames.

The wax glands are located under the mirrors and secrete an oily liquid which diffuses through the membrane at a temperature of 95°-97° F. and sets into scales of wax. The actual wax fragment, finally removed, often consists of two or more separate laminations, with each layer usually displaced slightly from that below, indicating that after one wax scale has hardened it is lifted by pressure from the next forming beneath it.

The scales are 'water white', i.e. clear, transparent and only appearing otherwise when so roughened that light is scattered on the surface to give an apparent whiteness. Indeed, freshly made comb, especially when made rapidly in a honey flow, also has this

17

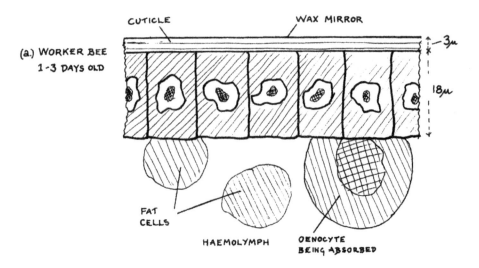

CUTICLE   WAX MIRROR

(a.) WORKER BEE
1-3 DAYS OLD

3μ

18μ

FAT
CELLS

HAEMOLYMPH   OENOCYTE
BEING ABSORBED

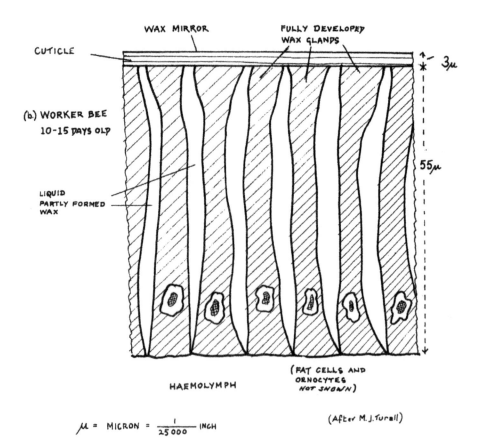

WAX MIRROR   FULLY DEVELOPED
WAX GLANDS

CUTICLE

3μ

(b.) WORKER BEE
10-15 DAYS OLD

55μ

LIQUID
PARTLY FORMED
WAX

HAEMOLYMPH

(FAT CELLS AND
OENOCYTES
NOT SHOWN)

μ = MICRON = $\frac{1}{25000}$ INCH

(After M. J. Turell)

FIG. 3.   Wax glands—detail

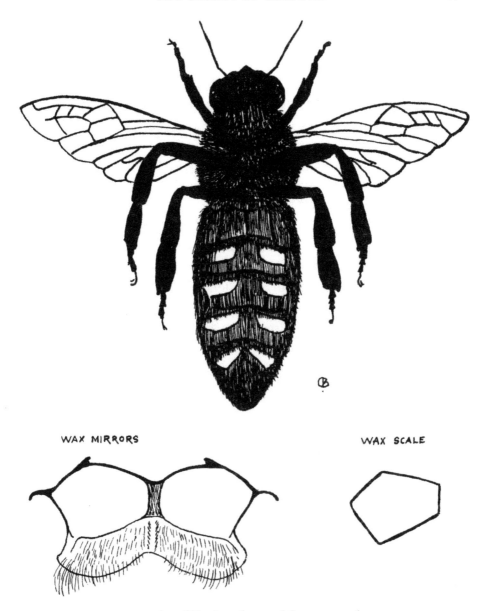

WAX MIRRORS

WAX SCALE

FIG. 4. Worker bee with wax scales

'rough-cast white' appearance, and any colour is attributable to later influences such as the thin varnish of propolis and saliva with which each inner cell surface is painted, or the bright colours derived from pollen fats (especially from dandelion pollen, for example, which even colours the woodwork of clean, new hive frames, walked over by bees which have been visiting dandelion).

Newly hived swarms usually drop large numbers of wax scales on the floor, and if a sheet of paper is inserted for an hour and then withdrawn, a sample of wax scales can readily be obtained. If some of these are gently warmed in a watch glass over a spirit lamp, a clear colourless liquid is obtained, whereas an equal quantity of new white wax from virgin comb only a day old, similarly melted, gives a small coloured residue. This suggests that bees moulding and working wax add something to it, and the colour probably comes from flavones in propolis or lipids in pollen, and is not inherent in the wax itself.

Cowan (like Huber) says that wax scales are removed from the mould by means of the pincers on the hind legs, but it is now known that this is not so. W. Herrod-Hempsall in *Beekeeping, New and Old* first gave a careful and detailed account of the actual transfer, and described how a scale is impaled on the downward-pointing stiff hairs (spikes almost) on the basitarsus of a bee's hind leg. From there the scale is either grasped direct by the mandibles, or more often transferred rapidly by the middle pair of legs to the mouth. The actual transfer action involving the middle pair of legs is so rapid that it resembles a conjuring trick and has to be watched several times before the human eye sees just how it is done.

WAX PRODUCTION

The actual production of beeswax takes place when young bees are gorged with honey or sugar syrup and hang together in a cluster clinging to each others legs by their claws, maintaining a temperature of $95°$-$97°$ F. After 24 hours the first platelets of wax appear, and the process goes on so long as fresh supplies of honey are coming in. Bees distinguish clearly between current revenue and capital, wax being produced on the former and not from capped stores also available in the hive. That is why it is so important to feed slowly and steadily a nucleus which has to draw out comb from foundation, or a swarm newly hived, above all when there is only a weak or intermittent honey flow.

The precise nature of wax production remains a bio-chemical mystery, but is attributed to enzyme action on sugars. As long ago as 1793 Francis Huber showed by careful experiment that bees made wax when fed on either honey or sugar syrup, but not when fed on pollen alone. It has long been a matter for controversy as to how much honey is used up by bees in manufacturing a pound of wax, and estimates given by various books range from 5 or 6 up to 20 lb. In his original experiments Huber found that bees confined to a hive and well fed produced a pound of wax for every six pounds of honey consumed, and went on doing this for seven days successively (the same bees). He then repeated the experiment with sugar syrup and found that just over 5 lb sugar (reduced to syrup) were needed per pound of wax produced.

Simmins (1886) carried out a very careful series of controlled experiments in which measured allowances were made for the honey or sugar syrup used to keep a similar number of bees and queen alive. He found that it took $6\frac{1}{4}$ lb honey to produce 1 lb beeswax, and conjectured that in more favourable circumstances (warmer weather) the figures might be as low as $5\frac{1}{2}$ lb.

WAX BUILDING

A worker bee exuding wax scales will normally disengage herself from the cluster and go to the nearest point where wax is required, cling on firmly to the work point, impale a scale on basitarsal spikes and transfer it via middle legs to mandibles. Should the work point be a cell under construction, the bee will hang on to the cell with four legs, and will use the front pair to help control the scale being held and 'worked' by the mandibles. The scale is gripped so that an edge is presented between the mandibles, and can then be revolved so that in turn every part can be chewed, saliva and glandular secretions added, and the wax worked until it is pliable, when it is placed in position and trowelled by the mandibles to the correct thickness and shape. Should there be no cells and no foundation, the wax-working bee will go to the nearest high point, possibly the branch from which the swarm is hanging, or the top of a hive frame, and there hang before chewing and depositing her load. The first few wax loads will be applied in irregular lumps and used as secure footholds, then subsequent wax loads will be deposited to form a more or less continuous ridge, which will be thickened by more loads until the depth is sufficient to be sculpted into the first cell bases. If the cluster of wax builders is relatively

small, then the first ridge may only be an inch long, and from it
will depend a curved tongue of honeycomb, being extended to each
side and downwards as successive workers come forward with more
wax.

Wax softens at a temperature of around 90°F (32°C) and except
on some days in high summer it is usually necessary for a cluster
of bees to form wherever wax has to be worked. It is interesting
to note that cappings of cells containing honey or pollen are made
of pure beeswax, but brood comb cappings contain about 90%
wax and almost 10% of pollen grains, plus fragments of old
cocoons. This permits slow ventilation of brood cells, the entry
of oxygen and the exit of carbon dioxide. Formic acid, one of the
medicaments used against the varroa mite, has a molecule only
marginally larger than a molecule of carbon dioxide, and so easily
passes through the porous brood cappings. No other anti-varroa
treatment is able to do so.

Throughout this book we are speaking of our own honey bees,
apis mellifera, but there are other bees also living in colonies
throughout the year (mostly in the tropics), and using wax of
almost identical composition.

They are:

(small)	*Apis cerana* (Indian bee) 80 cells per sq. inch.
(very small)	*Apis florea* (Arabian bee) 170 cells per sq. inch.
(large)	*Apis dorsata* (Eastern bee) 40 cells per sq. inch.
(for comparison)	*Apis mellifera* (ours) 50 cells per sq. inch. (25 on each side)

CHAPTER THREE

WAX PRODUCTION AT HOME

SAVE WAX SCRAPS

Apart from the main source of wax, that obtained from cappings at extracting time, there are opportunities for gathering scraps of wax throughout the year; from scrapings when frame tops in the hive are given their spring clean, from odd pieces of burr comb found linking two frames on a routine inspection, from that piece of wild comb built down from the crownboard when a frame was accidentally left out. The best solution is to form the habit of saving every fragment of wax, however small, by having a couple of 2 litre (or half-gallon) plastic ice-cream containers, with the usual clip-on lid, clearly labelled 'BEESWAX' with a black spirit marker, and keep one always in the shed or workshop and one in car or bee-bag.

USING A SOLAR EXTRACTOR

Most of the wax described so far will be suitable for melting down in a solar wax extractor, and it is very convenient to have one standing in the garden from April to September, always ready to receive the contents of one of the wax boxes. Against the rain, a water-proof cover can easily be made from an old plastic fertiliser bag cut down one side and roughly stitched to make a square or rectangular lid, fitting like a deep roof does on a hive. The point here is that it is so much easier to remove a light, plastic cover than to open up shed or garage and carry out a solar wax extractor. Keep it in a sunny corner of the garden all the summer. Purpose made solar extractors can readily be bought, but it is not difficult to make one at home.

HOW A SOLAR EXTRACTOR WORKS

If we first consider the basic principles involved, perhaps the logic of the construction features will be more apparent. The reader may remember from school physics that heat travels in three ways, by conduction, convection and radiation. Our problem is to get the heat into the box and keep it there, so the first essential is obviously a transparent top to let the sun's rays through as completely as possible. Fortunately glass, and most plastics, allow short-wave solar radiation to pass through, but are opaque to long-wave radiation of heat from the warmed up interior of the box. This is the so-called 'greenhouse effect'. We still have to stop heat loss from conduction, and this is why all conventional designs of solar extractors are double glazed, with a half inch air space between. But glass is expensive as well as fragile, and the writer found by experiment some years ago that one sheet of *thick* plastic was as good an insulator as double glazing. This is because plastic is a much better insulator than glass of the same thickness.

In making an efficient solar extractor, the other large surface is the base, and this must be a good insulator. This is best achieved by having a floor of 3/8″ wood or exterior $\frac{1}{4}$″ ply for weather resistance, then a layer of glass wool (Superwrap or some such proprietary brand as used in roof insulation), or a sheet of white polystyrene foam, perhaps obtained for nothing from old packing material. Over this it is advisable to have a thin sheet of ply to protect the insulation.

The side walls should be made of at least 3/8″ timber, and it probably isn't worth the trouble of having any extra insulation here, although it would help. The last point is that the box should be carefully made so that no air can leak out via cracks or poorly fitting corners. Any air leaking out will carry a good deal of heat with it, and if amateur carpentry leaves obvious chinks then some putty or filler compound must be used. The exterior woodwork of the solar extractor should be painted white, and the interior a dull black. The box has now been made very easy for solar heat to enter, and very difficult for heat to escape. It still has to be adapted for its purpose, i.e. melting wax.

Inside the box one needs a sloping surface made of fairly heavy metal, some arrangement for holding back the dross or slum-gum and allowing only the molten wax to go through, and room to put in a container into which the molten wax may drip. Just what materials are used and how they are placed depends on what is available, but a very satisfactory arrangement can be made with scrap material readily found, such as a piece of galvanised iron roofing (corrugated), an old iron oven top or the metal from an old hive roof. To hold back the wax fragments it is good practice to fasten a long strip of diamond mesh extruded metal across the middle of the metal plate, in the shape of a very broad V. Suitable off-cuts of this material can be found on building sites, and are often used by beekeepers as mouse-excluders in winter. At the lower end, the metal plate should be bent upwards to form a shallow funnel about two inches across from which the wax will drip, and here a piece of fairly open wire gauze can be slotted in for the wax to filter through. As this will get blocked from time to time, it is convenient to have two gauze filters and to drive two pairs of nails or bolts through the metal to form slots, making it easy for the gauze filters to be removed and replaced after cleaning by heat, or by wire brush.

Another method of holding back the black dross or slum-gum is to stuff the wax and comb fragments into old tights or silk

PL. 2. Beeswax blocks, nightlights and candles

PL. 3.   Wax extractor at work (60–150° F) in seven minutes

stockings and put these in the solar, when the molten wax readily runs through leaving the sediment behind.

MAKING A SIMPLE SOLAR EXTRACTOR

Illustrated in Plates 2 & 3 is a very simple but extremely effective wax extractor, of a type first made up by the writer in 1970. On a really hot day in June or July a block of wax weighing 2 to 3 lb has often been obtained, and the record, for one day in the summer of 1976, was $4\frac{1}{4}$ lb. Some of the blocks obtained are shown in Plate 4. Their somewhat distorted shape is due to the fact that the wax was allowed to run into a plastic ice-cream container, which was bent by the heat of the extractor.

The main body of the extractor was made from an old W.B.C. lift, and here the built-in plinth provides a ready-made seating for the 3/16″ plastic sheet. As W.B.C. type hives come in slightly different sizes, it is suggested that you first obtain a sound lift, as thick and strong as possible, and then measure up the exact size of the plastic sheet required, and have it cut to size at a D.I.Y. shop. The thick base of the extractor shown was made out of old floor

PL. 4.   A year's crop of beeswax, 47 lb from 42 hives

boards from a demolished house, sawn to size and firmly nailed on. Inside the box, strips of white polystyrene foam were laid over the base and then a sheet of thin plywood tacked down over this.

To carry the metal sheet on which the wax will melt, two light battens may be nailed or screwed on opposite inside walls, one on each, leaving sufficient room at one end for the wax receptacle. Finally, as the extractor will be used propped up at an angle of 35–40 degrees while the wax receptacle must be more or less horizontal, it is useful to tack a strip of wood diagonally across the bottom end to make a rough platform for the wax vessel.

The inward-sloping sides of a W.B.C. lift also ensure that the area collecting the sun's rays is maximised, and when correctly positioned the four sides receive little if any sun and so are best left white or silvery with aluminium paint to reduce heat loss by radiation. The use of a simple rotating stand (see diagram) makes correct alignment at right angles to the sun's rays more easy. There is scope here for an ingenious engineer to add a computer controlled motor to rotate the extractor from east to west to follow the sun (or for a lazy beekeeper to sunbathe alongside and give it an occasional push!).

The metal sheet inside should be dark, as should the inside of the box, so as to absorb radiation rather than reflect it back again.

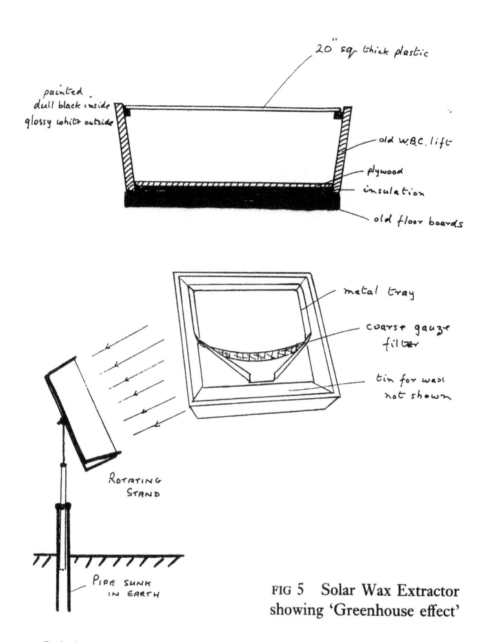

20" sq thick plastic

painted dull black inside
glossy white outside

old W.B.C. lift

plywood

insulation

old floor boards

metal tray

coarse gauze filter

tin for wax not shown

ROTATING STAND

PIPE SUNK IN EARTH

FIG 5   Solar Wax Extractor
showing 'Greenhouse effect'

It is important to appreciate the limitations of a solar extractor, which is very good indeed on odd combs and wax scraps, and excellent on dried cappings obtained when extracting. On the other hand it will not get any worthwhile amount of wax out of old, black combs. This is because the wax, on melting, is soaked up by the layer of old cocoons and pupa cases embedded in each cell.

Even so, it is convenient to put old brood frames into the solar for half an hour, when the hot comb can very readily be cut, or even pushed off the wooden frame, which is left clean and ready for the fitting of a new wax foundation. A great deal of the weight of these old brood combs is not wax at all, but layers of old pupa cases with debris at the base of each, plus propolis. W. F. Reid (1924) quoted actual wax content of some very old black brood combs as only 10% by weight, and some not quite so old as 15%. It takes a different technique, described later on, to handle combs like these.

Usually the bottom of the wax cake will be discoloured, for this is where the impurities settle. An effective way of dealing with this is to slice off the bottom to whatever depth is necessary, usually only about $\frac{1}{8}''$, and recycle this through the solar with the next load of wax.

One of the useful, and rather surprising, features of a solar wax extractor is the way that clean wax runs away from the slum-gum, which tends to get left on the metal sheet, and of course at the filters. The best way to clean this off is in the evening while the metal is still warm, using a flat paint scraper, saving the dark residue for processing with the older, black combs, in a manner now to be described.

REMOVING IMPURITIES

A reader of the first edition rang me up several years ago with an excellent idea. He remelts about $2\frac{1}{2}$lb of crude wax (unfiltered), either from his solar extractor or frame scrapings, and pours this into a two foot length of new grey plastic drain piping, lower end temporarily blocked with an empty plastic yoghurt carton used like a cork. When cool, the solidified wax shrinks away from the plastic and slides out very easily, with the dark impurities concentrated in the bottom three inches, which may easily be cut off and recycled, leaving pure wax above. A 2 ft length of piping 2 inches in diameter holds just over 2 pints, which if cast in a shallow mould would not give enough depth to isolate the impurities.

HANDLING OLD BLACK COMBS

Silk, whether from actual silk-worms or from the cocoon spun by bee larvae, has a great affinity for wax, and once wax has soaked into the silk cocoons no amount of boiling or pressing will get it all out again. On the other hand, if the cocoons are first soaked with water, they will not then absorb the wax.

So, it is very important to break up old black combs to expose as much surface as possible, and soak these in soft water for 24 hours before attempting to melt the wax out of them. Although there are purpose made wax extractors, it is possible to improvise, and one method is to enclose broken combs in a sack and boil up in an oil drum (previously cleaned) with soft water. Some wax will float to the surface, and much more can be made to do so by pressing hard on the sack (while still boiling) with a strong stick having a square of wood 3″ by 3″ screwed to the end. Left over night to cool, there will be a cake of solidified wax at the top, which can easily be cut round and removed.

OTHER TYPES OF WAX EXTRACTOR

There are two types of wax extractor normally sold for home use, one having a tall funnel into which water is poured to build up

PL. 5.   Wild combs

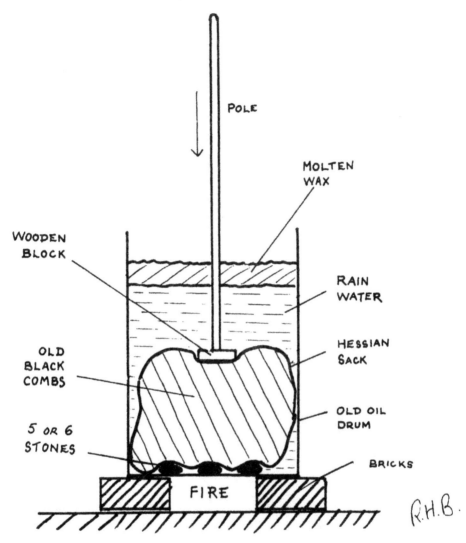

POLE

MOLTEN WAX

WOODEN BLOCK

RAIN WATER

OLD BLACK COMBS

HESSIAN SACK

OLD OIL DRUM

5 OR 6 STONES

BRICKS

FIRE

R.H.B.

FIG. 6.   Wax from old combs

hydraulic pressure and force molten wax through the filter (shown on page 32, Plate 6), and the other using steam. The former is really a modification of the 'boil in a sack' method, using water pressure to push the wax up through the filter. The latter employs a steam jacket to keep the old combs at a high temperature, while steam then circulates through the inner, perforated cylinder and helps the wax to run through and drip from a spout. Neither of these devices gets anything like all the wax out of old combs, as it takes a combination of steam and very high pressure to effect this.

PL. 6. M.G. wax extractor

FIG. 7. Coat of Arms of the Worshipful Company of Wax Chandlers

One small bonus of boiling up ones own old combs is that the black residues, when broken up, make excellent compost. The writer (like many beekeepers) is an enthusiastic believer in compost for gardens, and old comb residues rot down well with seaweed, dead leaves and garden refuse.

CLEANING WAX

Wax obtained from the solar by melting cappings and fairly new combs is usually of a good colour, but that obtained by boiling old combs is very dark and of limited use in that condition. Mr C. Brian Dennis has devised a technique for cleaning this wax and transforming it to a more acceptable golden yellow. In a large aluminium pan he uses about a pint of rainwater to every pound of dirty wax, adds about two teaspoons full of hydrogen peroxide and brings the mixture gently to the boil.

Although the peroxide may be responsible for some bleaching,

its main function is to produce a steady flow of tiny bubbles so that
steam may rise through the molten wax and wet the impurities
more thoroughly. Hydrogen peroxide readily gives off oxygen in
contact with small particles of organic origin, and these bubbles
also act as nuclei for steam to evaporate into, thus preventing
'bumping' or explosive boiling, which could otherwise be danger-
ous when handling hot wax. There is no point in using more than
the amount of hydrogen peroxide suggested, and any excess may
cause too much frothing and a boiling over of wax. After slow
cooling, with a folded newspaper over the pan, the impurities settle
to the bottom. The lower surface of the cake can be sliced or
scraped off to leave a good sample of wax. Even the scrapings
should not be discarded, but recycled, as they contain a good deal
of wax.

Mr. Dennis has found that beeswax is not appreciably affected
by being held at the boiling point of water for a considerable time,
so long as the water is soft (i.e. free from lime salts).

PL. 7.   Skep of natural combs

MANAGEMENT FOR WAX PRODUCTION

All the foregoing has been concerned with wax fragments or combs which arise naturally in the apiary, but it is possible to manage bees so that a larger crop of wax is produced. It is usually argued that comb building so diminishes the honey crop as to be totally un-economic, and therefore maximum yields are obtained by using supers of drawn comb and allowing brood combs to be used as long as possible. This ignores the fact that when a colony has a large proportion of young bees, these will often cluster and produce wax anyway, even dropping the wax scales on the floor. It is better to go with the bees and let them do what comes naturally—build comb in high summer. One very convenient way of doing this is to use wide-space drawn combs in supers by reducing from eleven frames to nine, or even eight, so that each year the bees can extend the cells to a greater depth. When uncapping at extracting time, cut deep down to the woodwork of the frames, and harvest the wax as well as the honey.

PL. 8.   Some good comb built naturally on starter strips of foundation

Again, the price paid for cut comb honey is at a substantial premium, so that it is worthwhile to go for a super of comb honey in a strong colony, using starter strips of foundation only $\frac{1}{2}''$ deep, so that there is no thick mid-rib in combs sold for the table. After cutting out 8 oz combs for plastic containers there is usually a good deal of partly sealed comb honey, or pieces of awkward shape, which go with the cappings and increase the yield of wax substantially. The writer kept a careful check several years ago on exactly what produce was yielded by 16 very ordinary supers of honey taken off at the end of the season. Apart from 395 lb of honey there were $9\frac{1}{2}$ lb wax, $1\frac{1}{2}$ lb pollen and $6\frac{1}{2}$ oz propolis.

It is usually stated that for every 100 lb honey, about 1 to $1\frac{1}{4}$ lb beeswax may be produced, but as demonstrated, this can be doubled without detriment to the honey crop. Indeed there is a bonus in that swarming tendencies are lower in stocks which have a normal amount of wax-building to do.

SELLING BEESWAX

The most efficient way, and certainly the most enjoyable, is at a bee stall dealing directly with the public, with live bees in a glass observation hive. The bee-populated combs provide a link with the beeswax exhibits. One ounce blocks of fine yellow wax sell themselves, and I have had customers coming back after 'impulse' buying to ask what they can do with the wax: once to ask if it should be eaten raw or cooked! Wax sewing buttons ($\frac{1}{3}$oz.) sell well at a bargain price of 10p, and one day I sold a 2 lb block of wax to a yachtsman about to sail around the world. Keen fishermen buy wax to use in tying their 'flies', as well as to rub on their lines. I recently ran a bee stall in the craft marquee at a church fete; on my left was a wood carver stabbing his specialised chisels into a much-used cake of beeswax (to make them grip the wood), and on my right a leather worker pulling linen thread across a wax nugget to make his sewing waterproof; both were supplied with new blocks of wax. Small buttons of beeswax ($\frac{1}{3}$oz.) were given to any pretty girl visiting the stall; the happy smile on the face of a nine year-old girl wearing a tooth-brace and receiving such homage was reward enough.

CHAPTER FOUR

THE USES OF BEESWAX

The uses of beeswax are so numerous and so diverse as almost to justify the production of a book on this aspect alone. They range from church candles to so extending the active life of penicillin in the human bloodstream that a single injection can cure gonorrhea: from waxing grandmothers' sewing thread to the manufacture of black camouflage face cream used in war by Commandos (and recently in London when S.A.S. troops rescued hostages in those dramatic scenes at the Iranian Embassy). In general terms, industrial nations tend to be net importers of wax, no matter how thriving their own beekeeping may be, and less developed countries tend to be exporters.

Looking at the overall world situation, there would appear to be five main fields of use, the most important single one being cosmetics, followed closely by church candles, with wax foundation for beekeepers and the pharmaceutical industry next in order. Most of the other applications would come under the heading of general industrial use.

COSMETICS

Basically, the use of beeswax in cosmetics depends on the saponification of the cerotic acid content of wax with a very mild alkali such as borax, plus the emulsifying action of a light oil like medicinal paraffin on the esters which make up the bulk of wax.

Details of the chemistry involved in this process are given in a technical appendix at the end of this book, but a basic recipe for cosmetic cream would be something like this:
15% beeswax (pure and of light colour), 55% liquid paraffin (medicinal), 29% water (rain or distilled), 1% borax (not boric acid), a drop or two of perfume.

The beeswax should be shredded finely and dissolved in the paraffin, heated to about 158° F (70° C) in a water bath, i.e. a double boiler, or a jug standing in nearly boiling water. The borax is separately dissolved in hot water at the same temperature and the wax/paraffin mixture poured into it while stirring vigorously at first, and occasionally while the temperature drops to about 122° F (50° C), at which the container feels quite hot, but may fairly comfortably be borne by the hand. Now the perfume is stirred in (it would have evaporated at a higher temperature), and when the mixture has cooled down to just over blood heat, it is poured into previously warmed (and dried) jars. This is not to prevent them cracking, but to stop any slight condensation on their cold inner surfaces, which could otherwise detract from the homogeneous appearance of the product.

A rather more solid cream may be obtained by slightly increasing the percentage of beeswax, and a lighter, 'fluffier' cream by slightly reducing the wax content. A cleansing cream would normally have rather less paraffin and rather more rain water, plus a little rose-water and maybe a cautious single drop of mild detergent, or not, as preferred.

A general formula for lipstick would be 35% beeswax and 65% castor oil (which is an excellent solvent for the dye used to give the colour), plus a trace of perfume if desired. Part of the beeswax could be replaced by lanolin.

Recipes are quoted in the appendix, but there is considerable room for variation to produce a product suited to the customers' requirements. What about a trace of royal jelly in the lipstick formula, for example?

CHURCH CANDLES

As will have been noted in the paragraphs on Wax Chandlers, in days gone by this was the principal use of beeswax, for general illumination and religious observance. In predominantly Roman Catholic countries this is still a very major use, although altar candles are no longer of pure beeswax as once they were. The actual amounts vary at the discretion of various Church authorities, and in Great Britain the content is now only 25%, but the Archidiocese of New York stipulates 51%.

Apart from the point of historical continuity, always important in church matters, there is a clearly defined symbolism which makes beeswax of supreme importance. Produced by virgin bees, the wax of the candle stands for the body of Christ, the Light of the World. The wick through the centre of the candle symbolises the soul of Christ, and the pure flame is the Holy Spirit, the divinity which dominates both body and soul. Just as the burning candle is consumed, so Christ died on the Cross.

Details of candle-making are given elsewhere, but certainly for church use the wax should be a yellow or golden colour, to show clearly that the candles are of beeswax. A very white or bleached wax would resemble too closely a cheaper, paraffin wax product. The wax must be pure, as the inclusion of even a small speck of propolis would show up as the light of the flame penetrates the translucent wax, and would also cause a noisy spluttering when it was drawn up into the wick.

In former days a large amount of pure beeswax was involved in the manufacture of Agnus Dei medallions. These were small discs of beeswax stamped with the representation of a lamb carrying either a cross or a banner (as the emblem of Christ) and the notation 'Agnus Dei' (Lamb of God). They were consecrated, or blessed, by the Pope and distributed after Mass at Easter. These medallions were greatly treasured and given a place of honour in Christian homes, in much the same way as the palm leaf crosses distibuted on Palm Sunday are today. The use of Agnus Dei wax medallions at Easter has greatly declined from former years and is now almost unknown.

WAX FOUNDATION

Many beekeepers feel strongly that this use should have priority, and that beeswax channelled into cosmetics and other uses is somehow wrong, in that it forces up the price of wax foundation. On the other hand a well-managed apiary, whether of two hives or, twenty or more, should consistently produce a surplus of beeswax over and above that recycled back into the business as foundation. The making of foundation, whether at home or in industry, is so important that a separate section has been devoted to it.

In recent years there has been a large growth in the production of candles made from rolled foundation. (See under Candle Making in Chapter 5.)

PHARMACEUTICAL

This covers a very wide field indeed, but mainly involves the use of wax in ointments, coating pills and in manufacturing processes.

One highly specialised use has already been mentioned, that of extending the viability of anti-biotics in the system. Normally penicillin in aqueous solution acts quickly and is soon eliminated, so that a number of doses at fairly short intervals are required, but it was found many years ago that injections of penicillin in a beeswax/peanut oil mixture had an extended life, giving a less violent but more sustained effect, thus permitting the use of larger doses at less frequent intervals without the risk of a violent reaction. The *Journal of the American Medical Association* on 22 January 1946 reported the persistence of an effective concentration of pen-icillin in the blood stream after 24 hours, and the use of 300,000 O.U. of calcium penicillate in a 4.8% mixture of beeswax/peanut oil with a complete absence of irritation. The point about the use of a single massive injection in the treatment of gonorrhea, is that patients most needing treatment often failed to turn up for extended courses, human frailty here involved in more than one sense. The *U.S. Army Medical Department Bulletin 81* of October 1944 reported 11 out of 12 patients cured of gonorrhea by a single injection of penicillin/beeswax/peanut oil complex. Beeswax is also used as a carrier for various ointments and emulsions, for coating pills and in suppositories. Dental wax, used to take impressions, is largely made of beeswax.

MADAME TUSSAUD

Born in Strasbourg in 1761, Marie spent her childhood in Paris, where her mother was housekeeper to Doctor Philippe Curtius. His natural interest in anatomy led him into the crafting of human figures in besswax and from him Marie learned the craft to which she was later to give her name.

At the age of 17 she was appointed art tutor to King Louis' sister and at the outbreak of the French Revolution in 1789 she returned to Paris, where later she helped Dr Curtius to mould the heads of guillotine victims, many known personally to her. What can have been her feelings, as she made Plaster of Paris death masks from the actual severed heads?

In 1802 she left her husband, Francois Tussaud, after a marriage which lasted eight years, and brought her children and her waxworks on a tour of England. Not until 1835 did she settle into a permanent home, in Baker Street, London, with her life-size exhibits of the famous and the notorious; from the Duke of Wellington to convicted murderers.

The waxwork of the lady herself was made in 1842, when she was 81 and said to resemble a waxwork anyway. She also acquired and exhibited authentic relics such as the original guillotine blade which severed the French royal heads and the carriage in which Napoleon escaped from Waterloo.

Four successive generations of Tussaud carried on the family business, located at its present site in Marylebone Road since 1884.

A few readers may also remember the great fire of March 1925, which demonstrated the inflammable nature of beeswax and destroyed the premises and most of the models, not long after my father had taken me to see this famous exhibition. A newspaper headline at the time read, 'Crippen Saved, Statesmen Perish'. The Chamber of Horrors survived, housed in the basement, together with many of the original death moulds made from guillotined heads. Within three years the show re-opened.

Towards the end of the nineteenth century some of Madame Tussaud's tableaux had become world-famous; for example: 'The Execution of Mary Queen of Scots' also, 'The Last Stand of General Gordon in Khartoum' which had a great influence on public opinion at that time and helped to change government policy.

Today, Warwick Castle also houses some historic tableaux, crafted in beeswax by Tussaud. They include an Edwardian Dinner Party, graced by the portly figure of King Edward VII himself, smoking a large cigar; also, the ladies, in their 'With-drawing Room' awaiting the return of their husbands; and a magnificent 'Musical Soiree' attended by the rich and famous.

CHAPTER FIVE

CANDLEMAKING

THE OLD DAYS

In Mediterranean countries, olive oil with floating wicks was used from the earliest times, and since olive oil will not rise very far up a wick, the lamps were flat, like slippers, as was Aladdin's magic lamp. Candles were used in Northern European countries where no natural oil was available. Poor people had 'farthing dips' or 'rushlights' made by dipping thin sedges or split rush stems in molten tallow (mutton fat) or beeswax. Wealthy people used beeswax candles for themselves and tallow candles for their servants, and for centuries both had to be 'snuffed' every half hour or so to prevent them from smoking. A split rush stem curved over naturally, as the hard outer edge became hot in the flame, but man was slow to learn from this. Not until a few years after the battle of Waterloo (1815) was it realised that a braided wick, with one thread drawn more tightly, would curve over naturally and burn away to keep itself the right length.

For many years the unit of illumination was, literally, a 'foot candle', defined as the light intensity produced by a 'standard candle' at a distance of one foot. The standard candle was defined as being made of spermaceti and burning at a steady rate of 120 gm per hour. This wax, obtained from the head of a whale, was regarded as a purer product than paraffin wax or beeswax. The term 'candle power' is still in use today, although it is now defined in a more scientific way. What a pity that a beeswax candle was not chosen!

BASIC THEORY

The job of the wick is to soak up the molten wax by capillarity so that it can be vapourised by the flame and burn away. If the wick is too thick for the candle, the liquid wax is soaked up too quickly, gives too rich a mixture and a smoky flame; also the wax is not melted fast enough and the candle flickers and almost goes out, then flares up again. In the wick is too thin, too much wax is melted for the wick to absorb, and molten wax runs down the side of the candle.

The actual candle flame has a small, cool centre usually blue in colour just above the wick, where the molten wax is evaporated into as yet unburnt gases, and a hot bright yellow/white top and outer edge where the gases are burnt completely away to give nothing but carbon dioxide and water vapour. If the wick stands up straight, it is in the cool part of the flame and does not burn away, getting longer as the wax is used up. This produces a smoky flame rich in soot (unburnt carbon particles), giving a dull yellow colour with less light. With a braided wick which curls over, the wick end itself burns away in the hot outer flame and stays the same size without having to be trimmed or snuffed. Candles can be made at home in five different ways, by rolling sheets of wax foundation, by the old Viking method, by pouring, dipping and using moulds. Whichever method is used, it is desirable to soak the wick first in hot wax 160°–170° F (71–77°C), spinning it between the fingers until all the air bubbles come out, and then drawing it out. After this wax has cooled, dip again and pull out at once, to thicken up the wax around the wick. Stretch the waxed wick taut until wax has set.

NIGHT LIGHTS

These were used many years ago to burn for an hour in a child's bedroom; when the child was asleep the light put itself out as it burned down to water level.* Slightly larger lights, also floating in water in decorative china dishes, give a soft illumination for intimate dinner parties.

* When children, my wife and her brother recall a poem in the bathroom attached to a night light saucer:

> When nights be dark,
> Then think of Clark
> Who does the job precisely
> For his night lights
> Produce light nights
> And show the way quite nicely

Editor.

Moulding these is very easy, as they are flat and a dozen at a time can be made in an ordinary domestic bun tin.

It is not necessary to go to the trouble of fixing the wicking in position before pouring the wax. As the depth is so small, a small hole can easily be made with a hot steel knitting needle and the wick threaded through. If the wick has been waxed and then slightly warmed, there will be no gap up which water can soak. If in any doubt about this, hold a lighted candle under the base for a few seconds after cutting the wick end flush.

The easiest way to melt the beeswax is in a double porringer or bain marie, and then pour it into a small, warm jug through a silk stocking stretched over a pulled-out metal coat-hanger. The small jug gives greater control when pouring from one mould to another around the baking tin.

ROLLED FOUNDATION CANDLES

In recent years candles, rolled from unwired wax foundation, have gained in popularity; with the attraction of their natural hexagon pattern and the ease with which they can be made. Also the fact that D.I.Y. foundation-makers are now much more common, apart from the facility of exchanging one's own beeswax for milled foundation offered by several firms.

For your first attempt, use a sheet of unwired shallow wax foundation (National) and a seven inch length of candle wick. The wax sheet can conveniently be warmed by blowing hot air over it from an ordinary hair-drier, or by placing it for about half a minute on a night storage heater at a low setting, with four to six thicknesses of newspaper on top to buffer the heat.

First, use your thumbs to roll over about $\frac{1}{4}$" of the narrow end, to form a U shape enclosing the wick, with an inch to spare at each end. You may have to blow warm air along the edge again first, but don't overdo it. Pull the wick taut and straight and gently compress the wax U to trap it. Now, roll up the wax sheet, keeping both ends level and when you get to the end, blow more more hot air along the join and press it to seal it down. The back of a warm teaspoon (or your thumb nail) can be used to 'iron it down'; when cool, it will be difficult even to notice the join. Trim the wick flush with one end and allow $\frac{1}{2}$" the other end for lighting. You now have a rigid candle of diameter almost one inch which will burn well and give a good light for almost three hours.

Another bonus with this method is that the wax is unlikely to be contaminated with propolis, slum-gum or moisture, which could cause spluttering and smoking if unfiltered wax had been used to pour into a mould. As with every other aspect of beekeeping, there is no substitute for 'hands-on experience' and it is suggested that you make just two candles like this at your first session and let one burn for twenty minutes or so, to see how it goes. Sometimes a candle may have been too loosely rolled, or the outer end of the wax sheet may not be well joined to the body of the candle, so be prepared to learn by experience.

Note carefully how the candle burns. If too much wax is melted, so that it runs down the side of the candle (called 'guttering') then the wick is too thin to take up and use the liquid wax fast enough. If the flame leaps up and then goes small before leaping up again ('flickering') the reverse is true. I mention this 'trial and error' approach, because much of the candle-wicking on sale is made for use with paraffin wax, which has a lower melting point (about 55°C, compared to about 63°C for beeswax). The so-called one inch cotton wicking, sold for cheaper paraffin wax candles of that diameter, is too thin for beeswax candles of the same size and I have found $1\frac{1}{2}$" wicking more suitable. I do realise that tubular wicking, especially for beeswax candles, can also be bought, but still recommend the 'try it and see' approach.

The next step forward is to wax the wick before use. This can be done very easily, by pulling a length of wicking taut between your two hands and running it horizontally through the pool of melted wax in a lighted candle. Hold it taut until the wax has set and you have a waxed wick much easier to put into your next candle, with an end much easier to light that a dry wick. It is important to use purpose-made braided cotton wicking, which bends over when lit and burns away in the oxidising area of the flame; thus, there is no need for 'snuffing' from time to time to avoid smoking. Supplies of wicking can be obtained from hobby shops and from bee appliance dealers; ordinary string just will not do.

If you now wish to vary the procedure by making tapered candles, cut the sheet of foundation diagonally in half before rolling. For a short taper at the top cut off a triangular strip an inch wide at one end. From now on the size, variety and decoration are up to you.

VIKING METHOD

Warm a long slab of wax in soft water at a temperature of about 120 F (49°C) for an hour until it is pliable. Place on a smooth, level surface and cut off four equal rectangular strips of wax. Lay two of these alongside each other and stretch the waxed wick down the centre division, then overlay the last two strips so that the wick is in the centre and roll the four together until round and smooth.

CANDLE-POURING

The first essential is to wax up the wick very thoroughly as previously described, but to get more wax on so that after rolling on a smooth surface the actual wick itself can hardly be seen. At this stage it is difficult to keep the wax temperature up, so a large container of warm, slightly soapy water at blood temperature or just below should be handy for dipping the young candle in and rolling (after wiping dry). Allow 2″ of wick to protrude free of wax.

Now have your supply of molten wax ready, also a ladle with a wide pouring lip. A support for one or more candles in the making can be contrived from a kitchen stool turned upside down, with 3 or 4 cuphooks screwed into the woodwork, and the pot of molten wax placed on the seat, directly under the wicks, each hanging from a small loop knotted in the free end. Before starting to pour, have an additional supply of molten wax keeping hot on a water bath, check that there are no draughts from open windows (to cause unequal cooling) and put a sheet of kitchen foil (aluminium) under the wax pot to catch the drops.

For the actual pouring, rotate a hanging wick so that it will spin back on release and gently pour a ladle of molten wax down the wick. The wax should be only a few degrees above melting point; if it is too hot it will melt some wax off the top of the young candle, and if it is almost setting it will not run down evenly.

Repeat with the other two or three hanging wicks, and after a couple of goes unhook each in turn and smooth down with a bare hand, or alternatively roll on a laminate working surface. By this method the candles may be built up to any desired thickness, but will be slightly narrower at the top. Subsequently the thickness can be evened up by dipping successively at different depths and rolling, if a truly cylindrical candle is desired. Finally cut off the 'icicles' of wax at the base with a warm knife and allow the candles to cool in a draught-free room, while suspended, before cutting the wick to size and moulding the candle ends on a hot surface.

CANDLE-DIPPING

The actual dipping pot must be deep enough to make candles of a reasonable length, but not too wide or it will take too much wax to fill it. It has to stand firmly in a much larger water bath, so that the water level is almost up to the wax level.

Clara Furness, in a memorable lecture at Seale Hayne College in Devon, demonstrated the use of fruit juice cans and catering size food tins soldered together to give adequate depth, and emphasised the need to weight down the wax can to hold it steady and prevent any tendency for it to float. There are some cylindrical tins deep enough themselves, such as Dr. Oliver's Bath Biscuit tins, and it is possible to buy specially made cans (address in appendix).

Remember that as the candle grows it displaces more wax, and allowance must be made for this when topping up the wax dipping can from the molten wax reservoir. Also the wax 'icicle' which forms at the base of the candle must be cut off from time to time, or it will make the growing candle too long to be immersed fully in the wax.

Although dipped candles were made several at a time commercially, the amateur will be wise to make just one at a time, at least to begin with while gaining experience.

Using a wick waxed as previously described, dip it gently full depth into the wax, hold for a second, and lift out. It may be necessary to straighten the young candle (until $\frac{1}{4}''$ thick) by rolling on a laminate surface. To begin with, the wax should be fairly hot 170°F (77°C), but once the candle is building up in size the temperature may be allowed to drop to 155°–160°F (68°–71°C) and then more wax will stay on the candle at each dip. A smoother finish will be gained, however, by using rather hotter wax for the final dip.

A 2" layer of molten wax on surface of hot water in a deep tin is sufficient.

The candle needs to cool for a few seconds, or longer, between each dip, depending on the air temperature. A cool room with no draught is best.

When the candle has reached the desired thickness, corresponding to the wick used, it needs to hang for several hours to cool and set completely firm; the final levelling off of the base can be done at any time with a warm knife. Dipped candles have a natural tapered shape, with a slightly wider base.

MOULDED CANDLES

Candles can be cast in a number of different types of mould, made of plastic, rubber, metal or glass, but basically the same rules need to be observed:

(a) warm the mould in a bucket of warm water containing a little detergent or liquid soap, (b) have the molten wax well above melting point, i.e. about 190° F (88° C), (c) after pouring allow it to cool slowly, wrapped in crumpled newspaper or glass wool.

Before pouring, the interior must be quickly dried with a well worn teacloth, and the mould (if rubber and so not rigid) supported in a large tin and packed round with sand to support it.

A large candle will take a long time to cool, and sometimes a hollow space may form at the top and go down some distance alongside the wick; this can be remedied by 'topping up' with molten wax, pushing down with a stiff wire while pulling up gently on the wick.

Wax shrinks as it cools and when it sets; with flexible rubber or plastic moulds the 'well' will not always form around the wick where it can be topped up, and the shrinkage elsewhere can distort the mould. To prevent this, do not allow the wax to set firm across the top of the mould.

Addresses of candle mould and wick suppliers are quoted in the technical appendix, but often old plastic containers can be used. The writer once made a perfect 2 lb beeswax candle in an old *Fairy Liquid* squeeze bottle, and burned it for dozens of hours during the power cuts of the late 1960's.

NEW TYPE WAX MOULDS

Made of a flexible, soft compound, usually about 5–10 mm thick, these are a great improvement on the older rubber moulds, which were often difficult to peel off candles or figures when there were details liable to break off. The modern versions also open up at one side, so that they can more easily be stripped from the casting with little risk. With candles, the wick is inserted through the centre of the base and secured by passing through the wickholder on top, allowing an inch to spare at each end.

If the purpose-made wickholder has been mislaid, then two cocktail sticks will serve, with a clothes peg gripping the wick above. The mould should be encircled with several rubber bands to hold it firmly closed. Pour the melted and filtered wax at a temperature not above 167°F (75°C) and be certain that the mould is completely cold before you remove the rubber bands and open up.

The thickness of the soft compound prevents over-rapid cooling and its flexibility takes account of the contraction of the wax as it sets. If there should be a hollow space as the wax contracts, it will be around the wick and if necessary, a little more wax can be poured in here.

MAKING MOULDS FOR WAX CASTING

Gelflex re-meltable compound has been designed for flexible mould-making. This substance, made from polymerised PVC resin, can be used to make moulds for candles of various shapes, small objects like teddy bears and rabbits, or even more easily of flat placques like the horse (see cover). A model of almost any rigid material can be used, provided that it does not soften at the melting point of Gelflex (120°–135°C). Porous materials should first be sealed with Unibond or some solvent based lacquer.

The model should be firmly seated on a base of laminate or smooth wood, with modelling clay or plasticine around the base

PL. 9a   Gelflex
moulding compound
ready for melting.

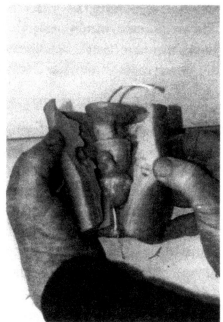

PL. 9b   Rabbit cast in beeswax
emerges from moulding.

PL. 9c   Candle made from a purchased mould.

to prevent leakage. Short lengths of grey plastic drain piping (off-cuts from a friendly plumber?) of a suitable diameter make ideal supports when forming the moulds.

Up to about 2.2 lb of Gelflex, cut into one-inch cubes, can be melted in a small saucepan with an asbestos or gauze mat over the hot-plate or gas ring. Start with about half the quantity you expect to use, stirring gently to keep the temperature even and prevent burning. When liquid, add the remainder, stirring occasionally. If over-heated, the material will rapidly change to a dark brown colour and may give off irritating fumes. Pour the molten Gelflex slowly down the side of the mould to prevent air bubbles and splashing; do not pour the liquid directly on to the model. Continue pouring until the model is covered to a depth of at least one inch above its highest point. Allow the Gelflex to cool and set before removing the mould from the model. With a flat object like the horse's head illustrated, the mould can easily be peeled off. If the model is larger and possibly has 'undercuts' or bulges, use a sharp knife to cut down the back of the model so that the mould can be opened up, like taking off an overcoat. Any small blisters or blemishes can be removed by careful use of a heated knife blade. The mould can be used for casting beeswax any number of times (the teddy bear mould illustrated had made over five dozen bears), but when no longer required it can be washed with water plus a little household detergent, allowed to dry and then cut up into pieces for re-use.

Gelflex moulds leave their castings easily and no release agent is needed. Care is always needed to avoid air bubbles, and of course any gaps extending from side to side or back to front may have to be plugged or it will be impossible to remove the mould.

Since molten Gelflex is about 30° hotter than boiling water, and will stick to the skin, it must be handled with great care, using gloves and eye protection against possible splashing. In the event of a skin burn, wash under running cold water or apply an anti-burn ointment with a loose dressing.

With normal handling it will not easily ignite, but in the event of a fire, use dry sand or a fire blanket.

Gelflex is obtainable from Trylon Ltd., Wollaston, Northants NN29 7QJ

POSTSCRIPT

For a joke, it is possible to bend a thin candle into a U-turn so that it can be burnt at both ends, to give point to the old verse:

> My candle burneth at both ends,
> It will not last the night.
> But ah my foes and oh my friends,
> It gives a lovely light.

(Edna St. Vincent Millay 1892–1950)

CHAPTER SIX

BEESWAX FOR HONEY SHOW

CHOICE OF WAX

Most successful exhibitors are agreed that the cleanest wax for show purposes is obtained from heather honey cappings or new comb built at the heather. If this is not available, then next best is any newly built comb and cappings from fairly new frames. Scrapings from crown boards, tops of frames and elsewhere provide good wax for general purposes, but not for showing. It is well-known that bees recycle wax, and if honey supers have been in use for some years, the wax in those combs is darker and when the bees cap the honey they often take wax from adjacent cells.

After draining and possibly centrifuging in a nylon wine filter bag in a spindrier, the comb fragments and cappings should be washed in rain water (or melted ice taken when a fridge or freezer is defrosted), drained and dried while spread out over a cotton sheet. When dry a scrutiny should be made and any darker wax pieces, pollen or flakes of propolis handpicked out. Wax contains about 14% of cerotic acids, so tap water, especially when hard (containing lime salts) must be avoided, as it reacts with the acid in the wax and forms a grey, mushy film. Rain water must not be used if from a roof with asbestos or cement; the writer's preference is for rain water collected from a greenhouse.

PRELIMINARY WORK

It is very convenient to use the domestic electric oven for the final pouring and slow cooling of the wax, last thing at night so there is no risk of anyone else switching the oven on to the wrong temperature (and less risk of the beekeper being tempted to look at the wax before it is thoroughly set). With this in mind it is wise to experiment with temperature settings, using a thermometer inside the actual oven, and to find the setting which gives a temperature range of 170°–180° F (77°–82° C), and mark this with a fine scratch on the oven knob.

A couple of clean builders' bricks, in a shallow baking tray placed on the top shelf, will act as a useful heat reservoir so that the wax cake make cool very slowly, and help to prevent premature cooling of the surface, which can result in a wavy appearance in the final product.

A *Pyrex* dish about 2″ deep with a diameter of about 6″ would hold about 1 lb of wax when half full; ideally it should have almost vertical sides so that the resulting wax cake will have no thin sharp edges which might be liable to break off.

Pyrex, *Phoenix* and other proprietary brands of ovenware are made from fused silica, which has a coefficient of expansion only one tenth that of glass. In other words it expands and contracts very little with change in temperature. This is an advantage as the wax then contracts on cooling far more than the vessel does, and so comes away more easily at the end. The depth to which wax has to be poured to provide a given weight can be determined by pouring in cold water while the mould is on the domestic scales, and marking the level on the outside with a spirit marking pen. The density of wax is only very slightly less than that of water, not enough to worry about in this connection.

A plastic bowl to contain warm water also has to be found for the mould plus wax to cool in. In order to have everything organised beforehand, it is wise to push the empty mould down in the plastic bowl and pour water in the bowl until it reaches a level slightly higher (say $\frac{1}{4}$″), than the level marked on the mould. By pouring off the water into a measuring vessel (or empty milk bottles), the exact amount of warm water required can be known beforehand. Have ready a sheet of glass for covering the mould while the wax is cooling in it.

The best filtering medium is ordinary white surgical lint (plain, not boracic), used with the fluffy side uppermost. Even better for the final filter is a large Whatman No. 1 filter paper (obtainable from chemists), but the pores are so small that only very clean wax will go through; wax that has not already been filtered through lint will contain tiny impurities which will speedily block the filter paper.

As a framework for filtering, i.e. to hold the lint on top and the filter paper beneath, the writer prefers a stiff loop bent from a wire coat-hanger, pushed into the nylon leg of an old pair of tights, and loosened to make a shallow bag. Alternatively one may use a clean tin, such as an old coffee tin with the bottom knocked out, also pushed into nylon mesh to support the actual filter and the weight of the molten wax.

MELTING AND WARMING

After the evening meal, put the tray of bricks on the top oven shelf and get them warm right through by heating the oven to about 200° F (92° C) for half an hour (unless it has been in use for a meal, when the bricks could have been on the top shelf then, and the oven setting turned back to the 170°–180° F (77°–81° C) when the meal was served). Now break up the selected wax into a quart glazed earthenware jug and stand this in a deep pan of slowly boiling water until the wax has melted. To prevent overheating the base of the jug, place an open ring of wire between it and the pan. Do not let the wax temperature go over 195° F (90° C).

Pour the wax when just above melting point.

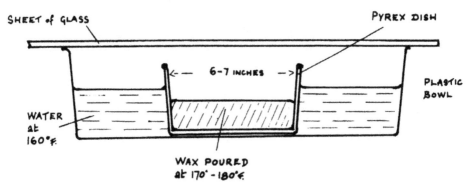

FIG. 9.   Wax cake cooling in oven

The mould, previously well washed and dried, should now be polished with the merest trace of detergent on a dry tea cloth, and placed upside down in the oven. Beside it place the jug of molten wax and the filtering apparatus. After ten minutes the wax jug will be quite dry and its top part and spout will be warm so that there will be no local cooling of the wax when the jug is tilted to pour it.

Some detergent molecules remain linked to glass or Pyrex even after drying and rubbing. These act as wax repellent and prevent sticking.

Half the correct quantity of water, at 160° F (71° C), should now be poured into the plastic bowl and the empty mould placed in it carefully to avoid any splashing; there will be no condensation as the surface is warm.

CASTING THE WAX

Now remove the jug of wax and the filter from the oven and pour the molten wax through the filter into the mould, from a close position to avoid splashing and minimise air bubbles; pour to the predetermined level and remove any air bubbles by touching with a warm knitting needle. Add the remaining hot water to the basin, replace in oven, and put the sheet of glass over the mould. Now switch off the oven and go to bed while the wax cools.

Next morning remove the glass sheet and take basin, water and mould out of the oven. Lift out the mould and invert it over a towel on a kitchen working surface: if the wax does not come out after a gentle tap, refill the plastic bowl with cold water and immerse the mould in it, so that water floods over the wax surface, when the wax cake should float up, and can then be inverted over a towel and gently dried, with no rubbing. If all has gone well the cake should not need polishing, except perhaps very gently with a piece of well worn silk, or with a dry thumb.

A JUDGE'S VIEWPOINT

For showing, there is nothing which sets off an exhibition cake of wax better than a piece of dark velvet, but a polished, glazed mahogany case is not a necessity, and it can be presented in a clean plastic bag, or on a plate.

The wax must be 'according to schedule', with a tolerance of 10% of required weight or thickness (B.B.K.A. Show Rule 10). It must be a flawlessly moulded, clean, bright, translucent sample of wax; above all no dark speck must show up when a torch is shone through the cake. Any colour between pale straw and light orange is accepted, but the main prize winners usually show wax of light primrose or golden yellow. A very light wax suggests bleaching, and a dark orange suggests rather too much dissolved resins from propolis. An uneven, wavy surface, or slight pitting from collapsed bubbles, will be penalised. Some judges will break off a tiny fragment to check texture and plasticity, possibly by chewing; any sign of adulteration will be heavily penalised. The aroma should be pleasant and suggestive of the kind of honey from which the wax was elaborated by the bees. Freshly moulded wax has more aroma than a cake several months old, and a very light specimen (possibly bleached) will have less aroma than a golden or primrose one.

PROBLEMS

Most of these arise from the fact that beeswax shrinks 9.6% as it changes from liquid to solid; in other words 1000 ml of liquid wax would give only 904 ml of solid wax. If the wax be poured into a cold container in a cool room, the wax on the outside rapidly solidifies and forms a thick skin around all the liquid wax, and as the remaining liquid contracts it either pulls away from the container, or the surface bends down in a curve. If the wax should stick firmly to the container and the surface layer is thick enough, large cracks will appear as the wax cools. Usually beekeepers warm the mould to slow down the cooling process, but forget to insulate the surface, which ends up with a pattern of wavy ripples on the surface. The point of slow cooling in an insulated oven, with a sheet of glass over the mould and warm bricks on the shelf above, is to keep the surface slightly warmer than the base and sides, to prevent this 'ripple' effect.

The other problem is when the perfectly formed wax cake sticks to the mould, and how exasperating this can be! It helps if the mould has a very smooth, highly polished surface. Also, as already mentioned, modern detergents have molecules which will to a certain extent 'bond' with a glass or Pyrex surface, so that some remain there even after the dry surface has been well rubbed with a clean tea cloth, and form an invisible 'non-stick' surface. This result is best achieved by standing the mould for 24 hours full of water with a squeeze of washing-up liquid, before pouring it out and polishing with a dry cloth. Do not touch any part of the clean inside surface with fingers after polishing.

CHAPTER SEVEN

MAKING WAX FOUNDATION AND CARE OF COMBS

One of the most important uses of beeswax is the production of sheets of wax foundation for beehives. Modern methods of management depend on the availability of wax sheets stamped with the hexagon imprint of cell bases, and about 25% of all beeswax is recycled back into hives as foundation.

THE WAX PRESS

The first home presses were made by Hertzog and imported from Germany, but in recent years they have been manufactured in Britain and indeed many amateurs now make their own.

The commonest type of wax press consists of a firm, tray-like base hinged to an upper surface which articulates exactly; upper and lower contacting surfaces are made of a metal alloy shaped in the exact form of the two surfaces of a sheet of foundation. Some of the home manufactured presses are made from fine plaster of Paris or other substances, but the technique is the same. The writer has over many years made hundreds of sheets of foundation, according to the following general method:—

Have ready a supply of beeswax melted over a low heat, preferably over hot water in a double saucepan, to a temperature of about 175°-185° F (about 83° C). Also a bucket of warm water (about 110°-120° F, 45° C) with a squirt of *liquid detergent* and the bubbles scummed off. Old newspapers are useful to spread over the working surface and floor, with half sheets of newspaper to interleaf the wax sheets on a flat surface.

First pour a breakfast cup full of the warm water into the press, lower and raise the top plate a couple of times and pour the water back into the bucket. Now scoop up a ladle (or half teacup full) of wax from the top of the pan, quickly pour it over the bottom plate in a Z-like motion and press the top plate down firmly; pour off any excess of wax at once, and within half a minute cut round the edge to remove the fillet of warm but solidified wax from the scuppers and open the press. Prise and peel off the sheet of foundation while still warm and lay immediately on a flat surface. Pour in another cup of warm water and repeat the process, establishing a rhythm which maintains the press and the water at a constant temperature. It is not necessary to use specially purified wax (except for show purposes) as any mucky bits are left at the bottom of the pan or ladle, never poured completely so as empty it.

If a large amount of wax is to be processed, for example by a group of beekeepers working together, it is an advantage to have two pots of melted wax, so that bits cut off, and additional wax, can be melted down to provide a continuous supply. In any case it is not practicable to add extra solid wax to the pan in use as it gets into the ladle. Usually the sheets produced are thicker than normal bought sheets and run out at just over five (National deep) to the pound. Using slightly warmer wax, and pressing harder, one may with practice achieve a steady six to the pound, seldom more.

Should the wax sheet stick in places, then gently prise out the fragments with a toothpick or matchstick, never with metal as the mould is relatively soft. If there are bare or very thin patches, more wax should be used at each pouring. It is much better to use an excess which runs off into the scuppers and can be poured off, than to use too little. Boiling or very hot water must *never* be poured over the press to clean it, as unequal expansion buckles the plate and ruins the press. It is very convenient to use the largest press available (M.D. size) and cut around a plywood template to reduce the wax sheet to National (or British Commercial) size.

The question is often asked 'What about disease transmission?' Research by more than one official body has shown that even wax melted from infected frames deliberately has failed to produce infection via wax sheets made from it.

THE HERRING FOUNDATION MAKER

The late Mr H. T. Herring of Basingstoke perfected a very different D.I.Y. technique, which first produces plain sheets of wax of any desired thickness, and then rolls them between strong

PL. 10. The author's home wax press

plastic dies to give the hexagonal cell imprint. Many readers have probably seen him demonstrating at the National and other Honey Shows throughout Britain.

Working with a supply of melted wax, either in a double sauce-pan or a deep 4 litre plastic ice-cream box standing in very hot water, a plastic jug is used to scoop up and pour wax into the tray, which should be held level, with the open end resting on the wax pot. Starting to pour at the closed end, the tray is flooded with wax and after about a second, is tipped to return the surplus wax to the pot. As soon as the wax has set sufficiently to be handled, he removes the wax sheet from the forming tray with the help of a knife and places it on the open dies. He closes the the dies at once, and starting from the hinge end, presses the hand roller down hard and rolls quickly down the centre, then down both sides. He opens the dies, and if he finds a part of the wax sheet without a clear pattern, closes the dies and rolls again while the wax is still soft. Carefully removing the wax from the dies, he places it on a flat surface, where it may be trimmed to the size required with a knife or trimmer wheel. This process produces more wax sheets per pound (and per hour!) than the wax press, and is more flexible in that thicker or thinner sheets of foundation can be produced at will.

Helpful hints

It is best to work in a warm room, temperature at least 70° F/21° C. Before use, spray the dies with hot water containing a few drops of washing-up liquid; in use the wax will keep the dies warm.

Do not allow your wax to get too hot: cool wax will lift off the forming-tray more easily than hot wax. A small teaspoon of honey added to the wax-pot will improve the quality of the foundation.

Any trace of wax sticking to the dies can be removed with hot soapy water and a brush. The wooden forming-tray should not be treated in any way or the surface will be affected. When not in use, store the dies flat and dry, away from hot sunlight.

Fill a bucket with luke-warm water and wet the forming-tray all over, allowing surplus water to return via the gap at the end of the tray, before starting.

FACTORY METHOD

Standard practice is to make huge rolls of plain beeswax about $\frac{1}{8}''$ thick by rotating a large, smooth, polished steel roller, cooled inside by water jets, dipping into a bath of melted beeswax. A knife edge placed in contact with the drum removes the wax in a continuous sheet which is automatically wound in a large roll.

In another machine, the wax ribbon is fed through a tank of warm, soft water and on between metal rollers surfaced with the hexagonal cell mould, to produce a continuous ribbon of foundation, cut to size and stacked automatically. There is considerable pressure between the rollers, which produces a sharp pattern very acceptable to the bees. Moreover, the actual pressure rolling affects the physical characteristics of the wax so that it is more flexible and less brittle than similar sheets made in a press. The distinction is analogous to that between malleable wrought iron and cast iron.

The writer enjoyed the privilege of seeing this process at the Wragby works of E. H. Thorne, and noted that wax from Ethiopia or Zambia imported via Dar es Salaam in huge dark cakes was blended with wax handed in by United Kingdom beekeepers to give a uniform golden yellow colour. It was interesting to learn

Thorne's beeswax room

PL. 11a.   Pouring hot wax

PL. 11b.   Making foundation

of the greatly increased demand for the old British Jumbo 14" deep sheet, also for sheets of drone foundation. The latter demand is due to the practice on the continent of putting a frame of drone comb alongside the brood nest and checking this frame for possible Varroa infestation, in view of the preference of this pest for breeding on drone larvae.

CARE OF WAX COMBS

Wax Moths

Almost every major book on beekeeping refers to the two species of wax moth: the greater (*Galleria mellonella*) and the lesser (*Achroia grisella*), and usually says that the former is the more important. In the writer's experience it is the smaller one which is found almost everywhere and does by far the greater amount of damage. In appearance it closely resembles its cousin, the common clothes moth, being about half an inch in length and usually of a drab greyish-brown colour with a silvery sheen. When disturbed it runs rapidly about on floorboard or comb surface, often with a deceptive zigzag avoiding action that makes it surprisingly difficult to catch; it is very flat in profile and able to squeeze through very narrow cracks. The writer's first acquaintance with wax moth was some forty years ago in Zambia, in his initial period of learning in isolation by trial and error. White, tapering larvae up to $\frac{3}{4}''$ long were seen on the hive alighting boards, being physically thrown out by the bees, and at first (in ignorance), thought to be bee larvae, until it was realised that they were too long and the wrong shape.

In the larval stage, the pest tunnels through wax combs, usually just to one side of the mid-rib, eating some wax but spoiling much more, protecting themselves by a web of spun silk reinforced by their excrement, making it very difficult for bees to get at them to throw them out. In extreme cases a box of combs can be reduced to a foul tangle of criss-crossed web tunnels with literally a thousand or more pupae lying flat on the frames or walls of the box. One surprising thing is the way the pest eats into the actual woodwork, making a shallow, tapering, oval depression in which the larva pupates, scarring the woodwork permanently.

Life cycle of wax moths

Although adult moths live only for a few days, long enough for the females to mate and lay up to 300 eggs (but usually less than 100), the life span of the larvae and pupae can be anything from a week or two up to several months, depending on temperature and availability of food. This means that the pest can overwinter in sheltered spots. When blowlamping (in April) hive floors not used for many months, the writer has frequently seen small wax moth larvae wriggling out of crevices as the flame drives them from cover. Their capacity for surviving many months in unused equipment is surprising, and seems to indicate a life-span rather longer than that usually quoted.

Winter storage of combs

The vulnerability of wax moths (eggs, larvae and pupae) to cold is the key to the problem, and even a couple of nights of frost with a cold day in between can kill them off. Research workers have shown that as little as 4 hours in a domestic deep freeze, long enough for the cold to penetrate through the combs, is enough to kill all stages of the pest.

For a beekeeper with just one or two hives, to freeze a super a night and then store them tightly wrapped in newspaper in a spare room might be a convenient solution. Just to wrap newspaper alone would invite disaster, which has been seen to happen in the case of beginners, who gained the impression from reading some bee-books that there was some protective virtue in newsprint. If eggs and larvae are already in the super, wrapping in newspaper is useless. The writer's practice is to build stacks of 8–10 supers with queen excluders or gauze screen boards at top and bottom, under a sound roof with reasonable ventilation. Two parallel angle irons resting on concrete blocks make a good support, keeping the base of the stacks 8–12″ off the ground.

If there is a queen excluder at the base, then the pile of boxes is almost certain to be occupied by one or two garden spiders, which will hunt down any moths present in the autumn (biological control!). On the other hand, inquisitive bees will be attracted and get trapped in the boxes on warm days in spring, especially if not all the boxes had been put back on hives after extraction for the bees to lick dry. Perhaps the soundest plan is to have a gauze screen board at base and a queen excluder at top (against entry by mice), and use a roof with an established family of spiders. Some authorities recommend stacking supers wet with honey, but this can result in water absorption, fermentation and a beery smell in spring; also some small crystals of granulated honey in the cells, which can then act as nuclei and induce early granulation of some honies in the combs when next used.

A beekeeper with a few hives might just as well winter a couple of supers on each hive, over the crown board with open feed-hole, so that bees can come up on mild days and look after their own combs. During a cold spell the bees will be clustered in the brood box and the temperature in the supers will fall below freezing and eliminate wax moth. Often one or two garden spiders will establish themselves in the roof, and should be tolerated; they will not eat more than half a dozen bees all winter, and for this modest fee they will also take care of any wax moths in a mild winter.

If a beekeeper does not wish to store supers of drawn comb on hives, or stacked in the open, one alternative is to put a teaspoonful of paradichlorobenzene crystals (PDB) in each newspaper-wrapped super. In a stack of honey supers kept in a warm shed or garage, it would be good practice to put a sheet of newspaper and a spoonful of PDB crystals between each of the boxes. Before using the hives next summer, the supers should be well aired to get rid of the odour.

Although 80% acetic acid is usually thought of as a reagent to sterilise combs against Nosema, it is also most effective against all stages of wax moth, from egg to imago, and so routine stacking and treatment with acetic acid in autumn would certainly take care of any wax moth problem.

Precautions against wax moth
Bearing in mind the habit wax moth have of laying eggs in crevices where the bees cannot reach them, the first precaution is to reduce the available crevices; do not use frames with a saw-cut, split top bar — the moth love to lay in this particular crevice, which inevitably becomes a nursery for them.

Check that the floorboards have not shrunk to leave a gap

between them, and if they have, hammer in thin strips of wood to fill. Check that the brood boxes are well made with closely fitting sides and no cracks; although the bees will eventually fill these cracks with propolis, this may not be until the colony has built up to full strength, and the wax moth may by then have established itself.

In general terms, a strong colony will be able to look after itself, but a small swarm or nucleus not occupying all the combs may allow wax moths to get the upper hand. Finally, clean the floorboards every spring and autumn, and do not allow any wax refuse to accumulate.

Handling Combs and Wax Foundation

In cold weather, wax combs are very brittle and clumsy handling can easily cause damage. The same is true of sheets of wax foundation, especially those pressed, as opposed to rolled or milled under pressure. Home-pressed foundation can be wired into frames immediately, while still warm, or stored on a flat surface interleaved with paper and strengthened with wire embedded by an electric current.

Beginners often fail to fasten new sheets of foundation securely in the frames, and when the sheet slips down and bends over, the resulting tangle of drawn comb represents the loss of some valuable three of four sheets of foundation.

Beeswax represents a valuable asset to any beekeeper, and supers of well-drawn comb are part of the essential working capital which money cannot buy. Every effort must be made to avoid waste and destruction by pests, careless handling and any loss of good combs or stocks of wax.

Honey Supers

In a bad summer perhaps not all your supers of drawn comb will be put in hives, yet if left idle another full year the wax may become brittle and darker, certainly less attractive to the bees.

It is a good idea to bring such supers into use by clearing supers of honey down into them, one on each hive. Even at the end of summer the bees will do some work on the combs and by their presence freshen up the combs, to keep them in better condition.

Left on hives over winter the bees will occupy them in April and early May and restore them completely.

Mini-nucs like the Apidea, should first be cleared of bees and food in October by pulling out the sliding floor and uniting to a full colony over the feeding hole, through a small strip of newspaper. After clearing (maybe two weeks), remove, clean and place for 24 hours in a deep freeze. Then, with entrance closed, the mini-nuc(s) may safely be stored in garage or home with no fear of damage by wax moth.

TECHNICAL APPENDIX

WAX PRODUCTION AND CALCULATIONS

Accepting Cheshire's figures of one pound of wax producing 35,000 cells to store 22 lb honey, and reckoning 500,000 wax scales to the pound, one can form some idea of the production effort involved. If each wax-working bee produced one batch of 8 scales every 12 hours, then it would take about 10,000 bees three days to produce one pound of wax (equivalent to 7 National deep frames of comb) and to produce this they would need to consume at least six pounds of honey or syrup. An average swarm of 20,000 bees, weighing 4 to 5 lb, would not be carrying with them more than $1\frac{1}{2}$-$1\frac{3}{4}$ lb honey at the most, so do not expect too much of a swarm in poor weather. In good weather with a steady honey flow, this same swarm, with 10,000 foragers, 10,000 housebees and no brood commitment for the first few days, could easily draw out a full box of combs and store a food reserve in the first week. A swarm twice the size could do this and fill a super in the same time. In poor weather the paramount importance of feeding a swarm is evident.

Wax scales

Those from the lowest abdominal segment are smaller than the others, but on average each scale is about $\frac{1}{8}''$ across and about $1/250''$ thick. Often laminated scales are found, consisting of up to four or five scales on top of each other, which accounts for the wide differences between their average weights as found by different workers. Gwin quoted 450,000 scales per 1 lb wax (1936), Pedley in 1952 found over 600,000. Roche gave a figure of 560,000 in 1968 and Robinson (1972—U.S.A.) even more. A reasonable approximation would be 500,000 per lb. This figure is consistent with the dimensions quoted and the known density of beeswax.

Durability of beeswax

(a) Scientific tests were made a few years ago on two cakes of beeswax found in an old Viking ship (dated A.D. 800) excavated in Norway. S.G. of 0.962, M.Pt. 147° F (63° C), acid value, saponification value, iodine number were all perfectly normal. There was no perceptable oxidation or deterioration despite passage of nearly 1,200 years. (b) Mr. Case-Green, of the British Wax Refineries Co. at Redhill, has recently handled a consignment of beeswax recovered from a cargo ship torpedoed off the coast of Scotland in World War I. The explosion which sank the ship also drove some live .303 ammunition deep into the wax blocks. Despite this, and having been immersed in seawater for over 60 years, the wax was quite undamaged. From its appearance, smell and general characteristics, Mr. Case-Green was able to identify the wax as having come from Angola.

THE CHEMISTRY OF BEESWAX

In most textbooks of organic chemistry the oils, fats and waxes are usually treated together, as they all consist mainly of compounds of the higher fatty acids (palmitic, stearic, oleic, butyric, etc.) with either glycerol, in the case of oils and fats, or the higher primary alcohols (cetyl, myricyl, etc.) in the case of waxes. The

71

distinction between an oil and a fat is based on melting point; if the substance stays liquid when cooled to 68° F (20° C) it is an oil. If it is solid at 68° F or above, then it is a fat.

The older books spoke of beeswax as a mixture of cerotic acid and myricine, the former being soluble hot alcohol and crystallizing out from alcoholic solution in colourless crystals. More recent books mention myricil palmitate and alcohol soluble ceroleine, and some refer to a hydrocarbon content and the presence of esters of other fatty acids.

The development of new analytical techniques involving gas and liquid chromotography, combined with mass spectography, has quite revolutionised our knowledge during the past 10-15 years, and it is now known that the volatile part of beeswax alone contains at least 111 compounds, of which only 41 have been so far identified. From the non-volatile part of beeswax 74 major and 210 minor components have been detected, so the picture is now a very complicated one, and it would be pointless to mention more than a summary. One interesting feature is that the composition of beeswax is still apparently fairly constant, whatever country it comes from, although characteristics such as colour and aroma enable experts, like Mr. Case-Green; to state the country of origin on inspection.

At the Madrid Apimondia Symposium in 1974 a team of industrial chemists (Iberceras S.A.) gave a clear summary of the present position, based on gas, liquid and thin film chromatography and I.R. spectography, which is here reproduced with acknowledgement.

Beeswax analysis (density 0.959-0.975 at 60° F, 15° C)

70% to 72% esters, in order of importance myricil palmitate, laceril palmitate, myricil oleo-palmitate, myricil hydroxy-palmitate, myricil cerostate and others.

14% to 15% free ceric acids, including cerotic acid, montanic acid, melissic acid.

12% hydrocarbons, mostly saturated, in the range C_{15} to C_{31}, including pentacozane, heptacozane, nonacozane.

1% free alcohols, 1% water and minerals, 0.5% lactone, 0.3% of dyestuffs such as dihydroxyflavone.

Other workers have confirmed these findings fairly closely and given still more detail, for example locating the hydroxyl group on the penultimate carbon atom in the case of hydroxypalmitic acid. The hydrocarbon fraction is now known to be about two thirds saturated and in the range C_{23} to C_{31} (mostly C_{33}).

In a book written for beekeepers there would seem to be little point in carrying the account any further than this, but the point is made that because beeswax is a mixture it has an ill-defined melting point 143°-147° F (62°-64° C) and that only pure chemical compounds, consisting of identical molecules, have sharp and exact melting points. Similarly the temperature at which beeswax becomes plastic, i.e. easily worked, is in the range 90°-95° F (32°-35° C), and this only happens because beeswax is a mixture of a number of separate substances, with crystalline structure varying from one constituent to another according to temperature.

Put simply, one could say that beeswax is a mixture of about 70% esters (largely myrical palmitate) 15% cerotic acid, 12% hydrocarbons, with traces of water, higher alcohols, minerals, dyes, etc.

The changes in the physical properties of beeswax on rolling under high pressure, as opposed to melting and casting, are analogous to the difference between malleable wrought iron and the more brittle cast iron. How extraordinary it is that all these highly desirable physical qualities should be found in a natural compound produced by the bees themselves.

Saponification

The damage done to beeswax in contact with hard water, and the basic reaction between borax and beeswax in the manufacture of cold cream and cosmetics, can only be appreciated if the term 'saponification' is understood. Essentially it means 'turning into soap' and is the basic reaction when a fat is boiled with an alkali to produce glycerine plus soap, thus:

Glyceryl stearate + sodium hydroxide = glycerine + sodium stearate
(i.e. mutton fat) + caustic soda = glycerine + soap.

The fats and oils commonly used in soapmaking are stearates, oleates and palmitates of glycerine, and all occur as common, natural products.

A very similar reaction occurs when a free fatty acid is heated with caustic soda, except that instead of glycerine, some water is formed. Just as in elementary chemistry we learn that acid + base = salt + water, so stearic acid + sodium hydroxide = sodium stearate + water; sodium stearate is the chemical name for common soap.

Beeswax contains 14 to 16% free ceric acids, the main one being lignoceric acid, $C_{24}H_{49}COOH$; hard water contains bicarbonates of lime and magnesia, which are weak alkalis. When heated together, calcium or magnesium cerate (technically a soap) is formed, and the nature of the beeswax is changed as its composition is altered by the partial removal of the ceric acid content. This is why distilled water, rain water or at least very 'soft' water, should always be used when washing comb fragments, or melting wax in contact with water.

Saponification also occurs when making cold cream and other cosmetics, but in this case it is desired in order to change the nature of the beeswax.

Ordinary borax is sodium tetraborate, $Na_2B_4O_7.10H_2O$, and is the salt of a strong base (sodium hydroxide) and a weak acid (boric acid). In water it is hydrolysed to produce an alkaline reaction, and will react with the ceric acids in beeswax to produce a 'soap' of sodium cerate, which is intimately mixed with the other constituents of beeswax to produce a 'fluffed out', creamy mass of myricil palmitate, laceril palmitate and many other esters. (An ester is the organic equivalent of a 'salt', i.e. the compound formed when an organic acid combines with an organic base.)

Just as sodium hydroxide (caustic soda) will saponify oils and fats to produce real soap, so borax (or potassium carbonate) will saponify the acid part of beeswax, but very much more gently, resulting in a product which is technically a 'soap', and indeed is sometimes used as one, e.g. cleansing cream.

The fatty acids in one gram of beeswax are neutralised by 0.068 gm of borax, so roughly 1 lb of beeswax needs at the most, slightly less than 1 oz of borax. Usually less borax than this is desirable, as any excess of borax may crystallise out and impart a rough texture to the product, the very last thing required in any cosmetic. This reaction is fundamental to the production of face cream, hand cream, etc.

Solubility of borax

Borax is not very soluble in water. At room temperature 100 gm of water will dissolve only 4 gm (Kaye and Laby), so that there must be 25 times the weight of water to borax to keep it in solution. The addition of a very small amount of glycerine increases the solubility considerably, and may be used without harming the final product.

Acid number

This is defined as the number of milligrams of pure potassium hydroxide required to neutralise completely the free fatty acids in one gram of the wax (or fat or oil).

Thus an average sample of beeswax, with an acid number of 20, needs 20 mg of potassium hydroxide per gram of wax to neutralise the acid content. The equivalent amount of any other alkali can be readily found by comparing their chemical equivalents; for example 68 mg of borax are equivalent to 20 mg of potassium hydroxide, so the acids in an average sample of beeswax (acid number 20) would need 0.068 gm borax per gram of wax.

Iodine value

This is defined as the amount of iodine which a wax (or fat) will absorb and chemically combine with, expressed as a percentage of the molecular weight of the wax. Saturated fatty acids absorb no iodine; thus the iodine value is a measure of the proportion of unsaturated fatty acids present. Vegetable fats normally contain a much higher proportion of unsaturated fatty acids, and much is made of this nowadays in dietary work when cholesterol build-up is of concern.

DETECTION OF ADULTERATED BEESWAX

1. *Specific gravity.* This is remarkably constant at values between 0.960 and 0.968, with 0.963 the most common. As vegetable waxes likely to be added are all much denser, a simple test may be made at home by adding water to methylated spirit in a glass until a piece of clean, freshly broken, solid beeswax with no air bubbles in it or on it, just stays submerged with no tendency to rise or fall. Then the s.g. of the liquid may be checked with a wine hydrometer. The suspect wax will probably sink rapidly to the bottom of the test liquid.
    2. *Melting point.* Pure beeswax has an extended M.Pt. of between 143°-147° F (62°-64° C), which may readily be checked.
    3. *Other tests.* Drop a little wax on an iron shovel or a tin lid heated on a stove or fire to just below red heat: (*a*) any adulterated fat—acrid smell of burning fat, greasy black smoke; (*b*) pure wax—clean, fragrant smell and when fractured shows a fine-grained, dull surface (any added resin—more polished look); (*c*) gives glossy surface when cut by knife; (*d*) warm wax in 10% H_2SO_4—resin gives red colour; (*e*) pure beeswax dissolves completely in CCl_4—impurities drop out; (*f*) force thumb-nail across wax surfaces—if good the thumb will pass jerkily; if thumb passes smoothly, likely to be adulterated, possibly with some fat; (*g*) push thumb-nail into wax; if pure a piece of up to 2 oz may be lifted up without falling off.

USES OF BEESWAX IN PHARMACY

Basic ointment carrier

Unguentum album of U.S. Pharmacopia consists of 5% bleached beeswax and 5% lanolin incorporated with 90% white petrolatum (petroleum jelly or vaseline). Used with various chemical additives for a great variety of conditions.
    The ability of beeswax to produce stable emulsions (both ionic and nonionic), is used in very many emulsified and dispersed pharmaceutical products, e.g. skin creams, sunproof creams, preparations for treatment of burns, etc.

Wax-oil fluid vectors

Many wax products are used as carriers, e.g. for penicillin as referred to in text. A Merck patent for a fluid oil/wax carrier has a beeswax content of 4.8%; *Ceratum camphore*—35% bleached beeswax; *Ceratum resinae*—20% yellow beeswax.
    Beeswax is soft, flexible, harmless, easily melted, insoluble in water but readily miscible with a large number of organic compounds. It is saponifiable and forms stable emulsions, altogether a near perfect vehicle for pharmaceutical products, from suppositories to lipsticks.
    Also used for coating pills to delay ingestion.

Oenocytes and fat cells

Young (1963) and Pick (1964) in a series of experiments fed or injected wax-working bees with radio-active C^{14} and deuterium and established that the acid and hydrocarbon fractions of wax were synthesised by the oenocytes and the esters by the fat cells. Both these types of cells discharge their contents into the wax glands.

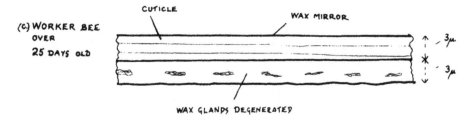

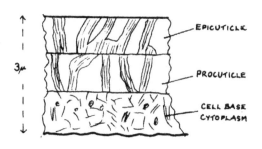

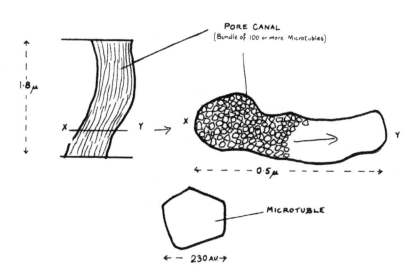

FIG. 10. Wax glands—fine structure

Wax mirror pores

Professor A. Dietz of Georgia University, U.S.A., published a paper in 1976 on 'The fine structure of wax glands', and found fairly conclusive evidence that the wax mirrors are traversed by minute pores through which wax flows outwards from the glands. Electron microscope photographs showed also that the diameter of these pores was only of one millionth of an inch.

Wax 'bloom'

The very thin layer of white coloration acquired by beeswax after several months is not due to oxidation or any chemical change, but a physical re-arrangement of molecules in the wax. This layer has been tested and found to have a slightly lower melting point than normal wax, but after melting with wax reverts completely leaving no residue of impurity. Mr. Case-Green, with a lifetime of experience handling beeswax, told the writer recently that,he has come to associate the 'bloom' on wax as an indication of high quality and purity.

Bumblebee wax

Secreted by queens as well as workers in the form of a greasy scum rather than as well-defined scales. Has a melting point about 77° F (25° C), lower than wax from honey bees, and is usually much mixed with pollen.

According to Dr. D. V. Alford, bumblebee wax and beeswax are very different in chemical composition and physical properties.

Central African beeswax

Obtained by squeezing honey from combs, melting down in old four-gallon paraffin tins, sometimes filtered through rush filters, sometimes not. At times brought to market as just cakes of pressed combs stuck together (sometimes with stones in middle!). Obtained from bark and trunk hives hauled up into trees, or from wild colonies robbed out. Melted into large blocks before export.

Lost wax process

A technique for casting metals, in which the object is modelled in clay, then thickly coated with wax, finally with an outer layer of clay, supported by a framework if large. On firing in a kiln the wax melts and runs out and a hollow space is left into which molten bronze, etc., can be poured. Described by Benvenuto Cellini of Florence in his *Autobiography, 1500-1571*, with many fascinating comments on the exorbitant cost of beeswax (and on the wayward nature of his patrons and fellow citizens). Cellini worked for several years under the patronage of Duke Alessandro in Florence, for whom he cast the famous statue of Perseus, the largest and perhaps the most difficult of all his work produced by the lost-wax process. In his autobiography he describes the stage by stage sculpting of the figure, made of clay on a frame of iron, coating it with beeswax and carving on it the fine detail. At this stage the observer would see exactly what the final work would be like, for the outer wax surface represented the appearance of the outer surface of the final bronze shell. Then the wax was covered with a 'tunic' of very fine clay, prepared months before and 'seasoned' to eliminate any large or gritty particles.

This outer clay layer was then reinforced with iron bands and given several vent holes at the base as well as an opening above. A slow fire was then built up to melt the wax and allow it to escape through the vents, (the 'lost' wax).

After this a funnel-shaped furnace of bricks was built around the mould of Perseus, the bricks arranged one above the other so as to leave numerous openings for the fire to breathe through. The fire was kept up for two days and nights on end so that the clay mould was baked hard. Then it was gently lowered into a large pit dug alongside and back filled with earth, to support the weight of bronze at the casting. Finally a furnace was constructed above the opening to the mould, plugged until the metal was molten. Cellini has described how he arranged the bars and pigs of bronze and copper so that the flames of the furnace would circulate, the lighting of the resinous pine logs, the throwing on of another 60 lb. of pewter, and in the end all the pewter plates and pots from his house in despair, lest there should be insufficient metal to fill the mould. By a miracle, when the plugs were drawn there was just enough metal to fill the mould with scarcely any left over. For two days Cellini let his work cool, and then uncovered it by breaking off the outer clay 'tunic' bit by bit. It was a great success and only the toes and a small piece of one foot had to added.

It is interesting to note that a similar process evolved quite independently in South America, though confined to rather smaller objects. The writer visited the 'Gold of El Dorado' exhibition, held at the Royal Academy in London from Nov 1978–March 1979, and the illustration shown (on page 43) is based on a hollow golden flask seen at this exhibition, on loan from the Museo del Oro, Bogota, Columbia.

OTHER WAXES

Carnauba has the highest melting point of them all, 181°-186° F (83°-86° C). Obtained from leaves of the Brazilian palm tree Copernicia cerifuga. The leaves are cropped from part of a tree twice a year, dried and beaten, when the waxy covering comes off as a fine dust, which is melted into cakes and exported.

Candelilla is obtained from a reed-like plant which grows in Mexico and California; has a M.Pt. of 158° F (70° C) sold in yellow opaque lumps.

Spermaceti, a fine wax obtained from head of sperm whale.

Esparto, obtained from esparto grass as a bye-product of the paper-making industry. M.Pt. 164° F (73° C), produces a high gloss with very little rubbing.

Ouricuri. Similar but not so good as carnauba; also obtained from leaves of a palm tree in tropical America. M.Pt. 183° F (84° C).

Shellac M.Pt. 166°-173° F (74°-78° C), excreted by the lac insect on trees in the Far East (like honeydew from aphids). Used for insulation in electrical industry and in polishes.

Sugarcane wax, dark in colour, M.Pt. 173°-176° F (78°-80° C). A by-product of sugar refining.

Paraffin Wax, the commonest of them all, obtained as a by-product of the oil industry. M.Pt. 130°-133°F (55°-56°C).

Ozokerite, an earth wax, mined like coal.

Ceresin, a mixture of purified ozokerite and paraffin wax.

WAX RECIPES – COSMETIC

Black mascara

Ingredients: 1½ oz beeswax, 1¼ oz carbon black, 4 oz triethanolamine, 5 oz paraffin wax, 1 oz lanoline.

Method: Warm and melt the ingredients and mill on a warm tile to ensure even mixing of the carbon black. Add a drop of oil soluble perfume, pour into shallow tins or jars and allow to set.

Cleansing cream

Ingredients: 1½ oz pure, light colour beeswax, 6 oz liquid paraffin, 4 oz rain water, ½ teaspoon borax, a few drops of rose water.

Method: Shred the beeswax and melt into the liquid paraffin at 160° F. Dissolve borax separately in rain water also at 160° F, and slowly stir this solution into the wax/paraffin mixture, taken off the heat. Add rose water and stir when cooled to a temperature the hand can stand. The addition of 3–5 drops of *Liquid detergent* adds to the efficiency of cleansing but must be avoided by those with possible allergy.

Cleansing cream (alternative)

Shred or grate 1 oz of beeswax into a double saucepan (porringer) or jug standing in hot water bath. Add $4\frac{1}{2}$ oz liquid paraffin, $\frac{1}{2}$ oz almond oil and 1 oz white vaseline (petroleum jelly) and bring up to a temperature of 120° F.

Dissolve $\frac{1}{2}$ small teaspoon ($1\frac{1}{2}$ g) of borax in 1 fl. oz of warm soft water and stir into the jug, having first taken it out of the water bath.

Eye shadow

Ingredients: $\frac{1}{4}$ oz beeswax, $\frac{1}{2}$ oz lanolin, $\frac{1}{2}$ oz paraffin wax, 1 oz liquid paraffin, $\frac{1}{2}$ oz white petrolatum, 1 oz blue/green or violet pigment.

Method: Melt all ingredients over gentle heat, mill while still warm and stir colour in well.

Face pack

Mix $\frac{1}{3}$ cup finely ground oatmeal with 3 teaspoons of liquid honey, blend to a smooth paste, add 1 teaspoon rose water. Apply evenly to face and leave for $\frac{1}{2}$ hour. Wash off with warm water and rub face vigorously with a warm towel before massaging gently with cold cream.

General-purpose cold cream

Ingredients: 6 pints liquid paraffin, $1\frac{1}{4}$ lb clean beeswax, 4 pints rain water, $1\frac{1}{2}$ oz borax, $\frac{1}{2}$ oz perfume.

Method: Shred beeswax and stir into paraffin at 160°F (71°C.) Dissolve borax into water also at 160°F, and pour into wax/paraffin mixture taken off heat and stood in sink of cold water. Stir as it cools, adding perfume at 140°F (60°C) and pour into jars at 120°F (50°C).

Hair cream

Ingredients: $1\frac{1}{2}$ oz beeswax, $\frac{1}{4}$ pint paraffin, 1 small teaspoon borax, 2 fl. oz rain water.

Method: Warm the liquid paraffin to about 160° F (71° C) on a water bath and stir in the shredded beeswax until liquid. Dissolve the borax separately in hot water (approx. $\frac{1}{3}$ teacup). Take paraffin and beeswax off heat and stir in borax solution, adding a trace of perfume if desired, as it cools.

Lipstick

Ingredients: 1 oz beeswax, 2 oz white petrolatum, $\frac{2}{3}$ oz liquid paraffin, 2 oz ceresin wax, $\frac{1}{3}$ oz cetyl alcohol, $\frac{1}{3}$ oz lanolin.

Method: Prepare a metal foil tube as a mould and invert into a sand-filled egg cup. Shred the waxes, heat very gently, stir well at the time, adding the other ingredients and finally the desired pigment.

Pour liquid into mould when almost cool, place into lipstick holder, then 'flame' it, i.e. pass it quickly through a low flame to give it a glossy finish.

WAX RECIPES—DOMESTIC

Green Mountain Salve (a very ancient remedy for rheumatism and back pain)

Melt slowly together 1 oz beeswax, 1 oz pitch, 1 oz mutton fat and 4 oz resin. Stir in 1 oz each of oil of red cedar, oil of turpentine and oil of wintergreen. Finally add $\frac{1}{4}$ oz very finely powdered verdigris. Roll into sticks when almost cool. Warm before use and rub well into affected area.

Furniture polish

Ingredients: 10 oz beeswax, 1 pint of natural turpentine. For a harder polish, replace 2 oz beeswax with 2 oz stearine or 1 oz carnauba wax.

Method: Shred the wax or waxes into a double saucepan over hot water and warm until all has melted. Stir in the turpentine, previously warmed with great care by standing in a container of hot water. Although very expensive, pure natural turpentine will make a fragrant polish which can be sold for use on valuable antiques. A much cheaper version may be made with the same volume of white spirit.

Furniture cream

Ingredients: 6 oz beeswax, $\frac{1}{2}$ pint rain water, 1 pint turps, $\frac{1}{4}$ oz borax (one teaspoon), $\frac{1}{2}$ oz soap flakes.

Method: Dissolve the borax and soap flakes in the hot rain water, add the shredded beeswax and warm until a fairly liquid mass, but fluffy. Now stir in the warmed turps and mix throughly. Pour into containers while still warm.

Floor polish

Ingredients: 3 oz beeswax, $4\frac{1}{2}$ oz paraffin wax, 1 pint white spirit.

Method: Shred the waxes into a small basin (enamel or stainless steel), standing in a sand tray (large tin lid partly filled with sand). Heat gently and stir well. Take away from stove for safety and after five minutes (to allow sand tray to cool to a safe temperature), add the white spirit gradually and stir as it cools.

Grafting wax (for gardeners)

Melt together equal parts of resin, beewax and lard. Roll out in sticks when almost cold.

Leather dubbin

Melt together 1 oz beeswax, 1 oz olive oil, 1 oz turpentine, 4 oz mutton fat, 2 oz lard. Stir thoroughly and pour into small jars or tins. Warm slightly before use and rub well in, to soften and waterproof new leather.

Liquid furniture polish

Shread 1 oz pure beeswax and shake in a bottle with $\frac{1}{2}$ pint carbon tetrachloride (dry-cleaning fluid, also used in some fire extinguishers). Wax is completely soluble in carbon tetrachloride and a few drops of the solution will give a thin film of wax over a large area with less rubbing needed. This is how aerosol polish is made.

Sealing wax (black)

Melt together and stir very thoroughly 1 oz beeswax, 1 oz lard, 3 oz lampblack, 15 oz resin. Pour on a smooth surface when part cooled and cut into smooth sticks.

Sealing wax (red)

As above, but substitute 1 oz or less red colouring matter (red ochre, red aniline dye, venetian red) for lampblack.

Sewing Wax

From ancient times, thread has been waxed to make it stiffer and to give better holding properties. The wax in a Viking ship dated A.D. 900 was found with a ball of thread, partly waxed, for sail repair. Experienced seamstresses and careful housewives like to have a small piece of beeswax in their sewing baskets, to draw thread over before use. The writer produces wax 'sewing buttons' by pouring molten bees wax into moulds, 24 on a sheet, made for home sweet production. These ¼ oz wax cakes sell at 10p a time, and often customers say they still have a similar lump their grandmother used. 'A life-time supply of wax for 10p', is a good selling line.

WAX RECIPES—MISCELLANEOUS

Antique furniture filler

Select 5–7 samples of beeswax to give a colour range from light yellow to dark brown. Take 5–7 ladles and melt equal parts of rosin and beeswax into each; hang on workshop wall. To use, match colour with job, choose correct ladle, warm gently and dig out filler as required. (Personal communication from antique dealer supplied with wax by writer.)

Brass-rubbing heelball

Melt together 8 oz beeswax and ¼ oz beef suet. When liquid stir in 2 oz ivory black or lamp black, ½ oz powdered gum arabic, ½ oz icing sugar and mix thoroughly. While still warm roll into balls of convenient size.

Crayons for drawing on glass or plastic

Melt together equal parts of beeswax and asphaltum, add a little lampblack and roll out in sticks on a smooth surface.

Drilling holes in steel

Warm the area of steel to be drilled, rub clean and then pour a little beeswax to cover the area and around. Within half a minute pour more wax on to thicken the cover. Make circular holes in the wax by hand spinning a drill of the right size; drop in a little concentrated nitric acid to fill the hole. Wash out after one hour and repeat until hole deep enough or right through. Finally pour boiling water to remove wax and last traces of acid, and smooth with small file.

Engraving metals

Warm the metal surface until when rubbed with a small piece of beeswax the latter melts and spreads over thinly; hold sideways so that excess wax drains off. Make the required drawing or inscription with a sharp point.

The engraving acid is made by mixing one part of concentrated hydrochloric acid with eight parts of concentrated nitric acid and applying with a glass rod. Within one to ten minutes, depending on the depth required, wash off the acid with water, warm the metal and remove wax by rubbing with rag. (Caution, both these acids are corrosive poisons and *must not* come into contact with flesh.)

Marble cement and filler

Melt together ½ oz beeswax, 4 oz resin, then stir in 2 oz plaster of Paris. Warm the marble and use cement while hot.

Painting (Encaustic): As at Pompei, painting with coloured beeswax on house walls – very durable.

Mix about 10% resin with beeswax melted in a waterbath. Make up a number of different colours with powder colour, a different one in each of 12 bun holes in a bun tray. Then by putting the tray on a hot plate (Low) all the wax colours melt easily and can be painted on a wall as a permanent, waterproof decoration.

(Source: an exhibit at the National Honey Show 1983)

Pipe threading

In screwed joints, the thread should be rubbed first with a little beeswax. This not only makes the joint completely waterproof, but also enables it to be unscrewed years later without risk of breakage.

Treating saws

Clean off rust with fine sandpaper. Stand in hot sun or heat gently in front of a fire, rub hard all over with piece of wax, and wipe off excess with an old rag. This prevents saw sticking even in green wood, and checks rusting. (Mr. John Starling of Borough Green, who taught writer to use a two-man cross-cut saw in 1934.)

Turners' cement

Melt together 2 oz beeswax, 1 oz resin, 1 oz pitch, and stir in 4 oz fine brickdust.

Varnish (general purpose)

Boil a pint of linseed oil with 1 oz beeswax until it goes 'stringy'; pour off the clear portion for use as varnish.

Waterproof Filler

Melt together 2 oz beeswax, 3 oz pitch, 8 oz resin. While still warm and flexible cut out and roll into convenient sticks. For use, melt only the quantity needed, as the filler becomes more brittle with repeated heating.

Wax varnish for statues

Melt together 1 oz white or very pale yellow beeswax with 4 oz turpentine. Brush on while still warm, having previously warmed the statue with cloths of hot water and wiped dry. Brush and wipe off varnish to leave thin film only.

CANDLE MAKING SUPPLIERS

E. H. Thorne, Beehive Works, Wragby, Lincs LN8 5LA (for wax and wicks)

National Bee Supplies, Okehampton, Devon EX20 1UD (for 1" and ½" wicks)

Camelot Country Products, High Leigh, Curry Rivel, Langport, Somerset
    TA10 0HB   www.honeyshop.co.uk

BIBLIOGRAPHY

Autobiography 1500-1571. Benvenuto Cellini

Observations on Bees. John Hunter. 1792

New Observations. Francis Huber. 1806

The British Bee Journal. 1863-85 (bound annual vols)

A Modern Bee Farm. S. Simmins. 1887

Bees and Beekeeping. F. R. Cheshire. 1892

Wax Craft. T. W. Cowan. 1908

Beekeeping in Antiquity. Malcolm Fraser. 1931

Beekeeping, New and Old. W. Herrod-Hempsall. 1937

Beeswax. H. R. Root. 1951

The Behaviour and Social Life of Honeybees. C. R. Ribbands. 1953

Beeswax. Edgar Pedley (Central Assn. Lecture)

Wax. H. A. Sturdy (Central Assn. Lecture)

The Wax Glands of the Honeybee. M. J. Turell (Master's thesis, Cornell University 1972)

The Wax Chandlers of London. John Dummelow. 1973

Composition and Analysis of Beeswax 'Apimondia Congress Madrid 1974' Iberceras paper.

Wax for Show. F. Padmore (National Honey Show leaflet)

Beeswax from the Apiary M.A.F.F. Adv. Leaflet 347

One Thousand Years of Devon Beekeeping. R. Brown. 1973

Bumblebees. D. V. Alford. 1975

Beeswax, Composition & Analysis. A. P. Tulloch (article in Bee World 61 (2): 47-62 (1980))

The Bee Book. Daphne More. 1976

The Fine Structure of the Wax Glands of the Honey Bee. Sanford M. T. and Dietz A. Univ. of Georgia, USA 'Apidologie' 1976 (3) 197-207

Beeswax. William Coggshall and Roger Morse. Wicwas Press, Ithaca, New York. 1984

Beekeeping—A Seasonal Guide. R. Brown. 1985

Honey Bees—A guide to management. R. Brown. 1988

INDEX

Lightning Source UK Ltd.
Milton Keynes UK
UKHW030645231020
372101UK00005B/240